Also by Lindsey Fitzharris

The Butchering Art: Joseph Lister's Quest to Transform the Grisly World of Victorian Medicine

THE
FACEMAKER

THE
FACEMAKER

ONE SURGEON'S BATTLE TO MEND
THE DISFIGURED SOLDIERS OF WORLD WAR I

LINDSEY FITZHARRIS

ALLEN LANE
an imprint of
PENGUIN BOOKS

ALLEN LANE

UK | USA | Canada | Ireland | Australia
India | New Zealand | South Africa

Penguin Books is part of the Penguin Random House group of companies
whose addresses can be found at global.penguinrandomhouse.com.

First published in the United States of America by Farrar, Straus and Giroux 2022
First published in Great Britain by Allen Lane 2022
005

Printed and bound in Great Britain by Clays Ltd, Elcograf S.p.A.

The authorized representative in the EEA is Penguin Random House Ireland,
Morrison Chambers, 32 Nassau Street, Dublin D02 YH68

A CIP catalogue record for this book is available from the British Library

ISBN: 978–0–241–38937–9

www.greenpenguin.co.uk

Penguin Random House is committed to a
sustainable future for our business, our readers
and our planet. This book is made from Forest
Stewardship Council® certified paper.

To my dad, Mike Fitzharris, who has always believed in me,
even when I did not believe in myself

He would show himself to the little guys and to their mothers and fathers and brothers and sisters and wives and sweethearts and grandmothers and grandfathers and he would have a sign over himself and the sign would say here is war and he would concentrate the whole war into such a small piece of meat and bone and hair that they would never forget it as long as they lived.

—Dalton Trumbo, *Johnny Got His Gun*

Only the dead have seen the end of war.

—George Santayana, 1922

CONTENTS

A NOTE TO THE READER

A significant challenge for any nonfiction writer is not to overwhelm the reader with too many details—something that is easily done when charting the immense scale of events that took place between 1914 and 1918. This book is by no means a definitive history of plastic surgery during the First World War. Nor is it a comprehensive biography of Harold Gillies, the surgeon who dedicated himself to rebuilding the faces of soldiers maimed during that time. For that, there are many articles and books written by scholars who have devoted their entire careers to these subjects, as my endnotes will attest. Rather, what follows is an intimate account of the daily struggles Gillies and his team faced at the Queen's Hospital, as well as the men who suffered the double trauma of injury on the battlefield and the painful process of recovery.

In their own time, disfigured soldiers were often hidden from public view. The decision to include their photographs in this book was not made lightly. I consulted various experts, including a disability activist with a facial disfigurement. The photos are undoubtedly graphic, and many people will find them difficult to view. But it is impossible to grasp the severity of these men's injuries and the reactions they elicited without seeing their faces. Equally, it is hard to appreciate fully the skill with which Harold Gillies and his team reconstructed soldiers' faces without seeing the surgical progress chronicled in these photographs. However, there is an exception: I have not included pre- or post-operative images of injured men who died in Gillies's care, as their reconstruction was never completed.

It bears stressing that this is a work of nonfiction. Anything placed between quotation marks comes from a historical document—be it a letter, diary, newspaper article, or surgical casebook. Any descriptive references to gestures, facial expressions, emotions, and the like are based on firsthand accounts.

It is my hope that through the telling of this tale, readers will gain a new perspective on the terrible consequences of trench warfare, and the private battles that many men fought long after they put down their rifles.

THE
FACEMAKER

PROLOGUE:
"AN UNLOVELY OBJECT"

NOVEMBER 20, 1917

Brilliant shards of crimson and gold pierced the eastern sky as dawn broke over Cambrai. The French city was a vital supply point for the German army positioned twenty-five miles from the Belgian border. On the dewy grass of a nearby hillside, Private Percy Clare of the 7th Battalion, East Surrey Regiment, was lying on his belly next to his commanding officer, awaiting the signal to advance.

Thirty minutes earlier, he had watched as hundreds of tanks rumbled over the soggy terrain toward the wire entanglement surrounding the German defense line. Under the cover of darkness, British troops had gained ground. But what had the appearance of a victory soon deteriorated into a hellish massacre for both sides. As Clare prepared himself for this dawn attack, he could already see the motionless, broken bodies of other soldiers scattered across the blasted landscape. "I rather wondered if I should even see another sun rise over the trenches," he later recorded in tightly lettered script in his diary.

The thirty-six-year-old soldier was no stranger to death. A year earlier, he had been holed up in the trenches of the Somme, where tedious stretches of inactivity were punctuated by frenzied bouts of terror. Every few days, wagons arrived to exchange rations for corpses. But the sheer number of bodies made it impossible to keep up. "They lay in trenches where they'd fallen," one soldier remembered. "Not only would you see them, but you'd be walking on them, slipping and sliding."

These rotting bodies became structural fixtures, lining trench walls and narrowing passageways. Arms and legs protruded out of the breastwork. Corpses were even used to fill in blasted roads that were essential for military vehicles. One man recalled that "they just shovelled everything into the crater and covered it over [with] dead horses, dead bodies . . . anything to fill up and cover it over and keep the traffic going." Common decencies were abandoned as burial parties tried to keep pace with the body count. The dead hung like laundry over barbed wire, covered inches deep with a black fur of flies. "The worst," remembered one infantryman, "was the bubbling mass of countless worms which oozed from the corpses."

The horror of these sights was exacerbated by the stench that accompanied them. The sickly-sweet scent of rotting flesh permeated the air for miles in all directions. A soldier could smell the front before he could see it. The stink clung to the stale bread he ate, the stagnant water he drank, the tattered uniform he wore. "Did you ever smell a dead mouse?" asked Lieutenant Robert C. Hoffman, a veteran of the First World War, when warning Americans against involvement in the second a little over two decades later. "This will give you about as much idea of what a group of long-dead soldiers smell like as will one grain of sand give you an idea of Atlantic City's beaches." Even after the dead were buried, Hoffman recalled, they "smelled so horribly that some of the officers became extremely sick."

Clare had grown accustomed to the dead, but not to the dying. The tremendous amount of suffering he had witnessed was etched into his mind. Once, he had stumbled upon two Germans cowering in a trench, their chests ripped open by shrapnel. The soldiers bore an uncanny resemblance to each other, leading Clare to conclude that they were father and son. The sight of their faces—"ghastly white, their features livid and quivering, their eyes so full of pain, horror and terror, perhaps each on account of the other"—haunted him. Clare had stood guard over the wounded men, hoping that medical assistance would arrive soon, but eventually he was forced to move on. Only later did he discover that a friend named Bean had thrust his bayonet into their bellies after Clare had quit the scene. "My indignation consumed me," Clare wrote in his diary. "I told him he would never survive this action; that I didn't believe God would suffer so cowardly and cruel a deed to go unpunished." Shortly afterward, Clare came upon his friend's decomposing remains in a trench.

Now, as he peered out over Cambrai's battlefield from his position on the hillside, Clare wondered what fresh horrors awaited him. In the distance, he could hear the faint staccato of the machine guns, and the whistle of shells as they sailed through the air. Clare wrote that upon impact, the "earth seemed to quake, at first with a jerk, like a giant startled out of sleep; afterwards with a continuous trembling communicated to us through our bodies lying there in contact with it." Shortly after the shelling began, his commanding officer gave the signal.

It was time.

Clare fixed his bayonet to his rifle and cautiously rose to his feet along with the other men in his platoon. He began marching down the exposed hillside. Along the way, he passed a stream of wounded men, their faces blanched with terror. Suddenly, a shell burst overhead, temporarily obscuring the scene with a cloud of smoke. Once it cleared, Clare saw that the platoon ahead of his own had been destroyed. "A few minutes later we moved on, stepping over the mutilated bodies of our poor comrades," he wrote. One corpse in particular drew his attention. It was a dead soldier who was entirely naked, "every stitch of clothing blown from the body . . . a curious effect of [a] high explosive burst."

Clare's own platoon continued to advance, passing through the carnage on the way to its intended target: a strongly fortified trench protected by a wide belt of barbed wire. As they drew closer, the Germans began raking them with bullets, their machine gunners and riflemen firing from several positions at once. Suddenly, Clare felt woefully underprepared. "[H]ow absurd it seemed to be advancing just one thin line of khaki, against the immensely strong entrenchment from which now belched a continuously increasing rifle fire."

Clare inched forward, weighed down by the heavy pack of supplies that all infantrymen were required to carry. These packs, which could weigh as much as sixty pounds, contained everything from ammunition and hand grenades to gas masks, goggles, shovels, and water. Clare negotiated tangles of barbed wire, keeping low to the ground to avoid the shower of bullets flying overhead.

Then, seven hundred yards from the trench, he felt a sharp blow to the side of his face. A single bullet had torn through both his cheeks. Blood cascaded from his mouth and nostrils, soaking the front of his uniform. Clare opened his mouth to scream, but no sound escaped. His face was too badly maimed to even arrange itself into a grimace of pain.

From the moment that the first machine gun rang out over the Western Front, one thing was clear: Europe's military technology had wildly surpassed its medical capabilities. Bullets tore through the air at terrifying speeds. Shells and mortar bombs exploded with a force that flung men around the battlefield like rag dolls. Ammunition containing magnesium fuses ignited when lodged in flesh. And a new threat, in the form of hot chunks of shrapnel, often covered in bacteria-laden mud, wrought terrible injuries on its victims. Bodies were battered, gouged, and hacked, but wounds to the face could be especially traumatic. Noses were blown off, jaws were shattered, tongues were torn out, and eyeballs were dislodged. In some cases, entire faces were obliterated. In the words of one battlefield nurse, "[T]he science of healing stood baffled before the science of destroying."

The nature of trench warfare led to high rates of facial injuries. Many combatants were shot in the face simply because they'd had no idea what to expect. "They seemed to think they could pop their heads up over a trench and move quickly enough to dodge the hail of machine-gun bullets," wrote one surgeon. Others, like Clare, sustained their injuries as they advanced across the battlefield. Men were maimed, burned, and gassed. Some were even kicked in the face by horses. Before the war was over, 280,000 men from France, Germany, and Britain alone would suffer some form of facial trauma. In addition to causing death and dismemberment, the war was also an efficient machine for producing millions of walking wounded.

The loss of life was also greater than in any previous war, due in part to the development of new technologies that enabled slaughter to occur on an industrial scale. Automatic weapons allowed soldiers to fire hundreds of rounds a minute at distant targets. Artillery became so advanced that some long-range weapons required their operators to take the curvature of the earth into consideration in order to remain accurate. The Germans' largest siege cannon, the dreaded "Paris Gun," pummeled the French capital with two-hundred-pound shells from a distance of seventy-five miles. Infantry weapons had also advanced considerably in the years leading up to the First World War, providing many times the rate of fire of those used in previous wars. The military historian Leo van Bergen notes that this, in combination with advances in artillery, meant that a company of just three hundred men in 1914 could "deploy firepower equivalent to that of the entire 60,000 strong army commanded by the Duke of Wellington at the Battle of Waterloo."

Beyond developments in the traditional hardware of guns, bullets, and shells were two ghastly innovations brought on by scientific advances. The first was the *Flammenwerfer*, or flamethrower, which produced an appalling shock for the uninitiated. It was first used by the Germans, most notably against the British at Hooge in 1915. The portable device belched forth a stream of burning oil that destroyed everything within range, sending men scurrying from the trenches like mice from burning haystacks. Its jets of liquid fire left victims with severe burns over their entire bodies. One soldier watched in horror as flames seared a fellow comrade: "his face [was] black and charred like a cinder and the upper part of his body scorched and cooked."

The second and perhaps more psychologically devastating innovation was chemical weapons. The first large-scale lethal gas attack came on April 22, 1915, when members of a special unit of the German army released 160 tons of chlorine gas over the battlefield at Ypres, in Belgium. Within minutes, over one thousand French and Algerian soldiers were killed, and a further four thousand wounded. Most of the survivors fled the battlefield with their lungs burning, leaving a large hole in the trench line. One soldier witnessed the horror from afar: "Then there staggered into our midst French soldiers, blinded, coughing, chests heaving, faces an ugly purple color, lips speechless with agony, and behind them in the gas-soaked trenches, we learned that they had left hundreds of dead and dying comrades." Even as gas masks were rushed to the front, offering varying degrees of protection, these chemical weapons became immediately synonymous with the savagery of World War I.

Tanks were also a new addition to the battlefield. First developed by the British, they were given their name in an attempt to conceal their true purpose from the enemy. Under the pretense of their being water tanks, these steel beasts were meant to protect those inside as they advanced their cannons and cargo inexorably toward enemy lines. In reality, they were vulnerable to shell fire, leaving their crews susceptible to all kinds of injuries, including burns from unprotected gas tanks that could ignite when hit.

Like Percy Clare, Captain Jono Wilson fought on the first day at Cambrai. He commanded a division of three tanks. Partway into his advance, Wilson's own tank ran out of fuel. He jumped out of the stalled vehicle, ran to the second tank in the formation, and climbed inside. Suddenly, that tank received a direct hit just as he was tying a message to a carrier pigeon. As the shell exploded, the vehicle toppled over onto its side, and fire erupted within. Before everyone could escape, the tank was hit again. The driver was killed, and Wilson's face was struck by white-hot shrapnel. While blood poured from the ragged crater where his nose had once been, he scrambled out of the tank and took cover in a shell hole, fortifying himself with a swig of rum from his canteen. He was eventually carried off the field by four German prisoners.

Meanwhile, in the skies above, pilots were engaged in dogfights or were taking fire from ground forces while flying reconnaissance missions. The planes—made of wood, wire, and canvas—were not bulletproof, and most airmen were just as vulnerable as their comrades on the ground. Air combat was in its infancy when the war began. It had been a little over a decade since the Wright brothers made the first successful powered flight, and airplanes were still primitive machines. Without parachutes, pilots were forced to crash-land burning aircraft or bail out and die. One pilot escaped with his body intact, but his face was so charred that none of his features was distinguishable. Most airmen carried a revolver or pistol, not to shoot the enemy but to end their own lives if their plane caught fire. So dangerous was flying in those days that many pilots were killed during training, before they ever had a chance to lay eyes on the enemy. These early airmen sometimes referred to themselves collectively as the "20-Minute Club"—the average time it took to shoot down a new pilot.

Yet for all these technological advancements, many of which were supposed to insulate the combatant from direct contact with the enemy, war was just as basic and brutal as it had been for centuries. Hand-to-hand combat broke out in scenes that would haunt survivors long after the war had ended. John Kirkham of the Manchester Battalion recalled the moment during the Battle of the Somme that he struck a German soldier with a trench club. This was a crude weapon, more redolent of medieval warfare than of the "modern" slaughter of the First World War. The standard-issue version was usually a kind of mace, or a lead-cored truncheon studded with hobnails, although they were sometimes improvised weapons cobbled together from various materials in the trenches. "It sank deep into his forehead," Kirkham recounted. "In the scuffle, his helmet flew off, and I saw that he was a bald-headed old man. I have never forgotten that bald head, and I don't suppose I ever will, poor devil."

Alongside the blunt clubs used in stealthy raids was the altogether sharper bayonet. None was more feared than the German sawback bayonet—nicknamed the "butcher's blade." Soldiers used its serrated edge to yank out the entrails of their enemies, causing slow and agonizing death for those on the receiving end. It was so loathed that the French and British armies warned the Germans that any man caught with one would be tortured and executed. By 1917, it had been widely outlawed in battle. But the invention and customization of weapons continued throughout the war, often with gruesome results.

Even discarded jam tins were made deadly early in the war as soldiers began improvising bombs by filling them with explosives and scrap iron and fitting them with fuses. Given the unprecedented proliferation of efficient ways to kill en masse, it is hardly surprising that the battlefield became a wasteland. In the words of one man, "there was not a sign of life of any sort . . . Not a tree, save a few dead stumps which looked strange in the moonlight at night. Not a bird; not even a rat, or a blade of grass . . . Death was written large everywhere."

These were just a handful of the horrors inflicted by the first of two global wars that would define the twentieth century. The conflict's human wreckage was inescapable. It was strewn across battlefields and crammed into makeshift hospitals all over Europe and beyond. Between eight and ten million soldiers died during the war, and over twice as many were wounded, often seriously. Many survived, only to be sent back into battle. Others were sent home with lasting disabilities. Those who sustained facial injuries—like Percy Clare—presented some of the greatest challenges to frontline medicine.

Unlike amputees, men whose facial features were disfigured were not necessarily celebrated as heroes. Whereas a missing leg might elicit sympathy and respect, a damaged face often caused feelings of revulsion and disgust. In newspapers of the time, maxillofacial wounds—injuries to the face and jaw—were portrayed as the worst of the worst, reflecting long-held prejudices against those with facial differences. The *Manchester Evening Chronicle* wrote that the disfigured soldier "knows that he can turn on to grieving relatives or to wondering, inquisitive strangers only a more or less repulsive mask where there was once a handsome or welcome face." Indeed, the historian Joanna Bourke has shown that "very severe facial disfigurement" was among the few injuries that the British War Office believed warranted a full pension, along with loss of multiple limbs, total paralysis, and "lunacy"—or shell shock, the mental disorder suffered by war-traumatized soldiers.

It's not surprising that disfigured soldiers were viewed differently from their comrades who sustained other types of injuries. For centuries, a marked face was interpreted as an outward sign of moral or intellectual degeneracy. People often associated facial irregularities with the devastating effects of disease, such as leprosy or syphilis, or with corporal punishment, wickedness, and sin. In fact, disfigurement carried with it such a stigma that French combatants who sustained such wounds during the Napoleonic Wars were sometimes killed by their comrades, who justified their actions with the rationalization that they were sparing these injured men from further misery. The misguided belief that disfigurement was "a fate worse than death" was still alive and well on the eve of the First World War.

A face is usually the first thing we notice about a person. It can signify gender, age, and ethnicity—all important components of an identity. It can also convey personality and help us communicate with one another. The infinite subtleties and variety of human expression comprise an emotional language of their own. So, when a face is obliterated, these key signifiers can disappear with it.

The importance of the face as a register of feelings or intent is even reflected in our language. We may attempt to "save face" or not to "lose face." If a person is trustworthy, their word can be taken at "face value." A liar might be considered "barefaced," "bald-faced," or even "two-faced." Someone might "cut off his nose to spite his face"—which brings to mind both metaphorical and literal disfigurement. The list goes on.

Disfigured soldiers often suffered self-imposed isolation from society following their return from war. The abrupt transformation from "typical" to "disfigured" was not only a shock to the patient, but also to his friends and family. Fiancées broke off engagements. Children fled at the sight of their fathers. One man recalled the time a doctor refused to look at him due to the severity of his wounds. He later remarked, "I supposed he [the doctor] thought it was only a matter of a few hours then I would pass out of existence." These reactions by outsiders could be painful. Robert Tait McKenzie, an inspector of convalescent hospitals for the Royal Army Medical Corps during the war, wrote that disfigured soldiers often became "victims of despondency, of melancholia, leading, in some cases, even to suicide."

These soldiers' lives were often left as shattered as their faces. Robbed of their very identities, such men came to symbolize the worst of a new, mechanized form of war. In France, they were called *les gueules cassées* (the broken faces), while in Germany they were commonly described as *das Gesichts entstellten* (twisted faces) or *Menschen ohne Gesicht* (men without faces). In Britain, they were known simply as the "Loneliest of Tommies"—the most tragic of all war victims—strangers even to themselves.

At Cambrai, Private Percy Clare was about to join their ranks.

After the bullet ripped through his face, Clare's first thought was that the wound was fatal. He wobbled on his feet for a moment before sinking to his knees, incredulous at the idea that he might die. "I had been through so many perilous times that I had unconsciously come to look upon myself as immune," he later recorded in his diary.

His mind began drifting to memories of his wife and child, when an officer named Rawson came to his aid. Shaken by the sight of Clare's ravaged face, Rawson tore out the packet of emergency field dressings that was sewn inside his own tunic. It contained lint, bandages, and a small bottle of iodine all rolled up in waterproof rubber. Rawson panicked when he was unable to determine the source of the bleeding and stuffed the entire packet into Clare's mouth before rushing back to the line to join his men. At that moment, Clare realized a man could easily drown from the torrent of blood caused by the rupture of major arteries in the face and neck. "Perhaps he . . . thought he could dam the outlet and thus stop the flow [of blood]," Clare later recalled. "As it was he only succeeded in nearly choking me, and I had hastily to gulp down the blood until I could snatch it out again."

Clare knew time was of the essence when his fingers began to tingle from the blood loss. He gathered what little strength he had and began crawling across the battlefield toward a road in the distance where he felt he had a better hope of being found. His limbs felt heavy, as though "a load of iron chains [were] about me," and he eventually collapsed before reaching his destination. There he lay, contemplating the nature of his own grave should he die: "I imagined the burial parties who perhaps tonight, perhaps tomorrow, would come along and find me, for this unsightly clay would be found eventually by strangers and buried in a shallow grave dug on the battlefield where I had fallen, as I, myself, had often buried others." He pulled a small Bible out of his pocket and clutched it to his chest, hoping that whoever found his body would post it back home to his mother.

As he drifted in and out of consciousness, he prayed that medical help would arrive soon. But Clare knew that the chances of a quick extraction from the battlefield were slim. Many men died waiting for the stretcher-bearers to arrive. A soldier named Ernest Wordsworth, who was injured in the first minutes of the first day of the Somme offensive, remained on the battlefield with blood streaming down his face for days before he was eventually rescued.

Encumbering the rescue process was the fact that stretcher-bearers couldn't step onto the battlefield without becoming targets themselves. During the Battle of Loos in the autumn of 1915, three men were killed and another four injured while trying to save a company commander named Samson, who had been shot just twenty yards from the trench. When a medical orderly finally reached him, Samson sent back a message that he was no longer worth saving. After the guns had quieted, his comrades found him dead, shot in seventeen places. His fist was jammed into his mouth so that his cries would not prompt any more men to risk their lives to save his. Tragic stories like this were far from uncommon.

Unsurprisingly, many soldiers died on the battlefield before ever receiving medical assistance. Attracting the attention of rescuers could be challenging, especially for those whose faces had been torn apart. The ghastliness of this type of injury could elicit terror in even the most battle-hardened warrior. The socialist activist Louis Barthas remembered the moment when one of his comrades was wounded. "We stood there a moment, horrified," he wrote. "[T]he man had almost no face left; a bullet had hit his mouth and exploded, blasting through his cheeks, shattering his jaws, ripping out his tongue, a bit of which was hanging down, and the blood gushed abundantly from these horrible wounds." The soldier was still alive, but no one in his squad recognized him without his face, prompting Barthas to wonder, "[W]ould even his own mother have recognized him in a state like that?"

In this respect, at least, Percy Clare was lucky. Despite the severity of his injury, he was still recognizable to a passing friend named Weyman. He heard a voice from above: "Hello, Perc, poor old fellow, how are you getting on?" Clare signaled with his hand that he thought the end was near. Weyman crouched down to assess the situation before alerting a stretcher-bearer. By then, the blood had started to congeal on Clare's hands and his face, even as it continued to trickle from the holes in his cheeks. The medical orderly just shook his head before ordering his men to push on. "[T]hat sort always dies soon," he muttered.

Weyman wasn't so easily deterred, however. He went in search of other stretcher-bearers as the shelling from enemy lines intensified. They, too, assumed Clare would die, and so they refused to carry him off the field. Clare was weakening by the minute and could hardly begrudge their decision. "I was so soaked with blood and looked so sorry a case that they probably were justified [in believing] that their long tramp . . . would be futile," he wrote.

To pick up a man like Clare, who seemed certain to die, meant leaving on the battlefield others with a better chance to survive, so decisions had to be considered carefully. Return journeys with the wounded were not only dangerous, but also physically taxing. Rescue equipment proved mostly useless in battle. Dogs trained to locate casualties were driven mad by shellfire. Wheeled carts designed to transport the injured were often useless on the blasted and furrowed ground. As a result, most stretcher-bearers had to carry men to safety with the stretcher on their shoulders. It sometimes took as many as eight people to move a single man. Nothing was easy, and nothing was quick. After Private W. Lugg picked up a man during the Battle of Passchendaele, it took him ten hours to travel through the mud before he reached help. Even when the extraction was a success, it was sometimes too little, too late. Jack Brown, a medic with the Royal Army Medical Corps, recalled that "it was then just a question of us lighting them a fag [cigarette] and saying a few words about the family at home until they died."

Given the location of his wound, Percy Clare faced another danger. Many soldiers who sustained facial injuries suffocated after they were placed flat on their backs. Blood and mucus blocked their airways, or their tongues slipped down their throats, choking them. One soldier recalled feeling a "smack" and then a dull thud as a bullet smashed through his face and lodged itself in his shoulder. "I was rendered speachless [sic] . . . My friends looked at me in horror and did not expect me to live many moments." They quickly bandaged his wounds but "were unable to stop the flow of blood in my mouth which was nearly choking me." He remained in the trenches, spitting up blood for hours, before finally being rescued.

Early in the war, the dental surgeon William Kelsey Fry discovered the challenges that facial injuries posed after he assisted a young man whose jaw had been blown apart during a night raid. Kelsey Fry instructed the soldier to lean his head forward to prevent his airway from becoming obstructed. After leading him through the trenches and into the hands of medics, Kelsey Fry turned around and began making his way back up the line. He hadn't gone fifty yards when a message was relayed to him that the soldier had already asphyxiated after being laid onto a stretcher. The experience stuck with Kelsey Fry for the rest of his life: "I well remember wrapping him in a blanket and burying him that night, and I made up my mind that if I had an opportunity of teaching that lesson to others, I would do so." Only later in the war did experienced medical officers like Kelsey Fry issue an official recommendation that soldiers with facial injuries be carried facedown with their head hanging over the end of the stretcher to avoid accidental suffocation.

In spite of all the daunting obstacles to rescue, Weyman was finally able to convince a third party of stretcher-bearers to take his friend off the field. Clare had lost a tremendous amount of blood by the time he was finally lifted onto a stretcher. He later referred to the wound in his diary as a "Blighty One"—demanding specialized treatment that would require his return to Britain, or "Old Blighty."

Any relief Clare might have felt at that moment, however, was short-lived. The next time he saw his face in a mirror, he received a shock. With a heavy heart, he concluded, "I was an unlovely object."

For Clare, the war might have been over, but the battle to recover had only just begun. Advances in transportation during the war had made it easier to remove injured soldiers quickly and efficiently from the battlefield. This, coupled with developments in wound management, meant that a large number of men were both *sustaining* and *surviving* injuries, including direct hits to the face. Improvements to sanitation within hospitals also meant that disease posed less of a threat to soldiers than in previous wars.

Injured men first received treatment at a regimental aid post, which was positioned immediately behind the fighting, in a relatively sheltered spot, or in a trench itself. They were then sent to a mobile medical unit known as a field ambulance, before being transported to a casualty clearing station a greater distance from the front. Although some casualty clearing stations were situated in permanent buildings—such as schools, convents, or factories—many consisted of large tented areas or wooden huts often covering half a square mile.

These facilities, which functioned as fully equipped hospitals, could be chaotic—especially at the start of the war. The British journalist Fritz August Voigt described one harrowing scene:

> The operating theatre looked like a butcher's shop. There were big pools and splashes of blood on the floor. Bits of flesh and skin and bone were littered everywhere. The gowns of the orderlies were stained and bespattered with blood and yellow picric acid [an antiseptic]. Each bucket was full of blood-sodden towels, splints, and bandages, with a foot, or a hand, or a severed knee joint overhanging the rim.

It was at a casualty clearing station that wounded men were stabilized and treated before being transferred by ambulance trains, road convoys, or canal barges to base hospitals along the French coast, some of which had as many as twenty-five hundred beds and were fully staffed with specialist doctors and nurses. Journeys to these facilities could take as long as two and a half days, depending on the mode of transport.

For soldiers who had received a "Blighty One," enormous hospital ships were on hand to shuttle them across the Channel to British ports. These ships were painted gray and bore large red crosses on each side to indicate that they were carrying wounded soldiers. Once they reached the other side, the men were transported to one of the many military hospitals that had been constructed during the war. Continuing improvements to this complex system led to a significant decrease in mortality rates over the course of the war.

Doctors and nurses at wartime hospitals experienced enormous challenges, but none was greater than the one posed by men with facial injuries. For them, survival alone wasn't enough. Further medical interventions would be needed to allow these men to return to some semblance of their former lives. Whereas a prosthetic limb did not necessarily have to resemble the arm or leg it was replacing, a face was a different matter. Any surgeon willing to take on the monumental task of reconstructing a soldier's face had to not only address loss of function, such as the ability to eat, but also consider aesthetics in order to reflect what society deemed acceptable.

Fortunately for Clare, a visionary surgeon named Harold Gillies had recently established the Queen's Hospital in Sidcup, England—one of the first in the world dedicated solely to facial reconstruction. Over the course of the war, Gillies would adapt and improve existing, rudimentary techniques of plastic surgery and develop entirely new ones. His unwavering dedication to this work was all in the cause of mending faces and spirits broken by the hell of the trenches. To help him with this daunting challenge, he would assemble a unique group of practitioners whose task would be to restore what had been torn apart, to re-create what had been destroyed. This multidisciplinary team would include surgeons, physicians, dentists, radiologists, artists, sculptors, mask-makers, and photographers—all of whom would assist in the reconstruction process from beginning to end. Under Gillies's leadership, the field of plastic surgery would evolve, and pioneering methods would become standardized as an obscure branch of medicine gained legitimacy and entered the modern era. It has flourished ever since, challenging the ways in which we understand ourselves and our identities through the reconstructive and aesthetic innovations of plastic surgeons the world over.

But on that late autumn morning in November 1917, Percy Clare just had to survive long enough to reach the medical help that he so desperately needed.

THE BALLERINA'S RUMP

The war and all its horrors were as yet unimaginable as Harold Delf Gillies and his wife wove their way through Covent Garden. Slender, with a beaklike nose and dark brown eyes that often glinted with mischief, the thirty-year-old surgeon had a habit of slouching that made him seem shorter than his five foot nine inches. The couple pushed on through the throng of stallholders and hawkers who were concluding their day's business on the cobbled streets. In the spring of 1913, London was a far more commanding presence in the world than it would be on the cusp of the Second World War, twenty-six years later. With over seven million people living there, this bustling metropolis was larger than the municipalities of Paris, Vienna, and St. Petersburg put together, and it was home to more people than Britain and Ireland's sixteen other largest cities combined.

London wasn't just big. It was also wealthy. The city funneled ships into and out of the North Sea via the Thames as they exported and imported goods from all points of the compass. It was one of the busiest and most prosperous ports in the world, and a vast emporium of luxuries. Dockers unloaded regular shipments of Chinese tea, African ivory, Indian spices, and Jamaican rum. With this influx of goods came people from countless nations, some of whom decided to settle in the capital permanently. As a result, London was more cosmopolitan than ever before.

Londoners worked hard and played harder. There were 6,566 licensed premises that fueled the city's favorite pastime—drinking—and ensured that the police force was kept busy. London boasted 5 football teams, 53 theaters, 51 music halls, and nearly 100 cinemas that would see weekly attendance triple by the end of the decade.

On that unseasonably warm spring evening, the Royal Opera House was staging its first performance of Verdi's *Aida* for London's more well-to-do music lovers. Gillies had been given tickets by his boss, Sir Milsom Rees, a laryngologist who specialized in illnesses and injuries of the larynx, or voice box. As medical consultant to the Royal Opera House, Rees was charged with tending to the throats of its famous singers. On this occasion, however, he was indisposed, so he sent his young protégé to deputize for him.

Three years earlier, Gillies had acquired his cushy position at Rees's medical practice, situated in the fashionable district of Marylebone, largely by happenstance. When he interviewed for the job, he had just completed his clinical studies at St. Bartholomew's Hospital in London. During that time, he had shown a keen interest in otorhinolaryngology, a surgical subspecialty that deals more broadly with conditions of the head and neck. Those who work in this field more commonly refer to it as ENT (ear, nose, and throat). The chief physician, Walter Langdon-Brown, considered him to be one of the ablest in his class. But it wasn't Gillies's surgical skills that had landed him the job with Rees across town. Rather, it was his reputation as an excellent golfer that had caught the older doctor's attention.

At the time, Gillies had just reached the fifth round of the

English Amateur Championship. During the job interview, Rees brought out his own golf clubs for Gillies to inspect. As the laryngologist demonstrated his swing, Gillies grew impatient. "This is ridiculous. When is he going to talk about the job?" he wondered. As it turned out, they never did find a chance to discuss the terms of employment. A short way into the interview, a patient arrived, prompting Rees to rush a bewildered Gillies from his office. Just as he was closing the door, Rees briefly swung his attention back to his would-be employee and offhandedly remarked, "Oh, my dear fellow, I'd forgotten! Well, how would five hundred [pounds] a year suit you? Any private patients you pick up you can keep for yourself. All right?" Gillies—who had been making fifty pounds a year at the hospital was elated at the prospect of making ten times as much money as an ENT specialist in Rees's private practice. It was not the last time that admiration for Gillies's sporting prowess would open the door to opportunity.

Gillies had always been a high achiever. He was a man for whom talent—be it athletic, artistic, or academic—was "mysteriously inherited rather than laboriously acquired," as his early biographer Reginald Pound observed. The youngest of eight children, Harold Gillies was born in Dunedin, New Zealand, on June 17, 1882. His grandfather John had immigrated there from the Scottish Isle of Bute in 1852, bringing his eldest son, Robert, along with him. Robert eventually set up business as a land surveyor, and it was in Dunedin that he met Emily Street, the woman who would become Harold's mother. The two fell in love and married shortly thereafter.

Gillies spent the first few years of his childhood tottering around the cavernous rooms of a Victorian villa. His father, an amateur astronomer, had commissioned the construction of an observatory with a revolving dome on the roof of their ornate stone residence. Robert Gillies christened the family home "Transit House." He chose the name in honor of the New Zealand astronomers who had made important observations of the 1874 transit of Venus when the planet passed across the face of the sun.

Gillies was a precocious child who loved to spend time roaming the expansive countryside around his home with his five older brothers, who would prop him up in the saddle of Brogo, the family mare, and bring him along on hunting and fishing expeditions. Early in life, Gillies fractured an elbow while sliding down the long banisters in the family home, which permanently restricted the range of motion of his right arm. It was a disability that later spurred him to invent an ergonomic needle-holder for use in the operating theater to compensate for his limited ability to rotate his hand.

Two days before his fourth birthday in June 1886, Gillies's idyllic childhood was shattered. That morning, one of his brothers climbed the stairs to check on their father, who had complained of feeling unwell the previous evening. When he entered the bedroom, he found Robert Gillies alert and in good spirits. His father told him that he would soon join everyone for breakfast in the dining room downstairs. The boy hurried off to tell his family the welcome news.

The kitchen sprang to life as pots and pans were pulled from high shelves, and the kettle whistled at the end of the water's slow boil. But as the minutes ticked by, Gillies's brother grew increasingly concerned. After half an hour, he climbed the grand staircase once more. A shock awaited him in the bedroom. Lying motionless on the mattress was Robert Gillies, dead from a sudden aneurysm at the age of fifty.

Following her husband's death, Gillies's mother moved herself and her eight children to Auckland so that they could be closer to her own family. When Gillies was eight years old, he was sent to England to attend Lindley Lodge, a boys' preparatory school near Rugby, in the heart of the country. Four years later, Gillies returned home to continue his education in New Zealand, but he wouldn't remain there for long. In 1900, at the age of eighteen, he moved back to England in order to study medicine at Cambridge University. His decision to become a doctor came as a surprise to everyone. It was a career he purported to have chosen to differentiate himself from his brothers, who were lawyers. "I thought another profession should be represented in the family," he joked.

At Cambridge, he gained a reputation for being something of a maverick after he spent his entire scholarship fund on a new motorcycle. He wasn't afraid to challenge his professors and could often be found arguing with the anatomical demonstrator in the university's dissection lab. Despite this lack of deference for authority, he was eminently likable and admired by teachers and classmates alike for "his happy temperament and his smile that broke into uproarious laughter." His popularity won him a nickname, "Giles," which stuck with him his entire life.

In spite of his rebellious spirit, Gillies had an orderly mind with an affinity for rules and boundaries—especially if he was the one setting them. For the duration of his studies, he lived in a Victorian terraced house with five other young men. As students are wont to do, they came and went as they pleased. Gillies noticed that not every housemate was present at mealtimes, so he devised a system to keep track of costs. Each person was required to mark down his attendance at meals in addition to the number of "units" he consumed, as well as the cost per unit. One of his fellow lodgers called it a "most original and ingenious scheme" that ensured equity and helped keep costs down for everyone. But his mates were less impressed when Gillies charged each of them interest on money that they owed him after he had settled a household debt. For Gillies, fairness was all.

It was during his studies that he developed a serious interest in golf, routinely swapping his pen for a hickory-shafted driver. He tried out for the university's golf team on a whim after traveling to Sandwich for a party with some classmates. He had brought his golf clubs with him so that he could play a round on the famous course there, where a match between Cambridge and Oxford was going to be held a few days later. After the party was over, Gillies boarded a return train. At the last second, he had a change of heart. He grabbed his clubs and hopped off the carriage just as the locomotive began steaming out of the station. Shortly afterward, he was welcomed onto Cambridge University's golf team.

Gillies spent an inordinate amount of time locked away in the bathroom, which must have raised a few eyebrows among his housemates. His daily ritual in the tiny room was to plant his feet on the same two patches of linoleum and practice his swing in front of the mirror. His friend Norman Jewson, who would later become a famous architect, was struck by Gillies's "immense powers of concentration, and will power." Those who knew him described his talent for golf as "supernatural." In time, his patients would come to see his skill as a plastic surgeon in a similar light.

As the years passed and his studies progressed, Gillies began to display an aptitude for surgery—which was not surprising, given his obsessive attention to detail. He was driven in a way many young men of his social class were not, often sequestering himself in a library while his peers were out socializing. One friend remarked, "Whatever he decided to do he did." His determination would serve him well in life.

This was never truer than when it came to matters of the heart. Although Gillies had vowed never to marry a nurse, he found himself suddenly and hopelessly in love with Kathleen Margaret Jackson, a nurse at St. Bartholomew's Hospital, where Gillies had been working during his clinical studies. But there was a problem: another doctor was also courting her.

Never one to shy away from a little competition, Gillies redoubled his efforts. One evening, he hired a hansom cab and invited Kathleen out for a ride. Once in the buggy, Gillies had the cabbie drive them continuously through the streets until she accepted his proposal. The stern etiquette of the day required that nurses live on hospital grounds and remain unmarried, so Kathleen resigned from her job shortly after becoming engaged. The two were happily married six months later on November 9, 1911. By then, Gillies was ensconced in his lucrative position at Rees's private medical practice.

It was with his wife, Kathleen, that Gillies was attending the performance of *Aida* in Covent Garden's grandly porticoed opera house on that pleasant spring night. The couple had left their firstborn— a little boy named John who would become a POW during the Second World War when his Spitfire was shot down over France—in the care of family. As the curtain fell on the opera's first act, a white-gloved attendant approached Gillies discreetly and requested his presence backstage. Given the habitually light duties of his boss on these occasions, Gillies expected to have to do little more than spray some sort of soothing balm into the overworked throat of a singer. Instead, he found one of the dancers injured and in a state of undress. Felyne Verbist, the Belgian prima ballerina, had sat on a pair of scissors, sustaining a deep puncture wound to her shapely backside. Gillies set to work bandaging the tender spot.

As he returned to his seat, he wondered how he would explain his prolonged absence—and the details of the "throat" case—to his young wife. Throughout the rest of the performance, he had trouble concentrating on anything "but the slight lump in the beautiful dancer's costume where my rather rough-and-ready dressing bulged."

It was an incident that Gillies would recount many times in later years, as if removing the pointed end of a pair of scissors from a ballerina's buttock was the crowning glory of his career.

Felyne Verbist was performing in the very same production of Aida *a year later on July 28, 1914, when the Austro-Hungarian Empire*

declared war on Serbia, signaling the start of the First World
War. A week later—as Britons flocked to the beach to enjoy one
last bank holiday before the summer officially drew to a close—
Britain declared war on Germany, plunging the nation into one of
the deadliest conflicts in history. On that sweltering summer day,
however, few people could have predicted the calamity that was
about to engulf the nation. The outbreak of war came as a complete surprise to most.

The trouble had begun a month earlier. A Serbian nationalist
named Gavrilo Princip had shot the Austrian archduke Franz Ferdinand and his wife, Sophie, Duchess of Hohenberg, while they
were visiting Sarajevo. The couple had traveled there to inspect
the imperial armed forces in Bosnia and Herzegovina, which had
been annexed by the Austro-Hungarian Empire in 1908. Princip
believed the territories belonged to Serbia and saw an opportunity
to retaliate for the annexation by assassinating the presumptive
heir to the imperial throne. Supplied with weapons by a Serbian
terrorist organization called the Black Hand, Princip and five other
conspirators met in Sarajevo with the intention of assassinating the
archduke.

Ferdinand was not oblivious to the danger. Three years earlier,
the Black Hand had tried to eliminate his uncle, the emperor Franz
Josef. And the archduke had allegedly told a family member shortly
before he died that he had foreseen his own murder. Nonetheless,
Ferdinand must not have been overly concerned for his safety on
that particular trip, since he announced his plans to visit Sarajevo
two months in advance of traveling—giving any would-be assassins plenty of time to formulate a plan.

In retrospect, it would seem that all the parties involved had a
date with destiny.

On the morning of June 28, the royal couple arrived by train.
They were in high spirits, as it was their wedding anniversary. Indeed, that was one of the reasons the duchess insisted on being

at her husband's side on this official state visit. Their personal chauffeur—a chubby-cheeked, neatly mustachioed man named Leopold Lojka—had accompanied them on their journey. Lojka helped the archduke and duchess into a Gräf & Stift Double Phaeton convertible with a license plate that read A111 118—a spooky coincidence, given that Armistice Day would later fall on 11–11–18.

The luxurious car was the second in a six-vehicle motorcade that was to proceed to the city hall along a tree-lined boulevard known as Appel Quay, which skirted the Miljacka River. The previous day had been cool and rainy, but the sun had broken through the clouds to welcome the royal couple on their visit. Given the glorious weather, the cloth top of the convertible had been folded down to allow people to see the archduke and duchess as they were driven to their destination. Official security precautions were conspicuously absent despite warnings that a terrorist attack was likely.

Armed with semiautomatic pistols and explosives strapped around their waists, the assassins had scattered themselves along the parade route earlier that morning to give themselves the best chance of intercepting the archduke. If one failed, another stood in reserve. In addition to their weapons, they also carried with them paper packets of cyanide powder, in case their plan went awry. It wasn't long before it did.

The first would-be assassin was a twenty-eight-year-old named Muhamed Mehmedbašić. As the motorcade rolled past him at a stately pace, however, he lost his nerve. He later claimed that a nearby police officer had spooked him, and he worried that he might put the entire mission in jeopardy if he failed to hit his target. Minutes later, the car approached Nedeljko Čabrinović, a nineteen-year-old who had a compelling reason not to fear the long-term repercussions of his actions: he was dying of tuberculosis—a condition that was incurable in 1914.

Čabrinović broke the detonator of a grenade against a lamppost and hurled it at the archduke's car. Lojka spotted the bomb flying through the air and slammed his foot down on the accelerator. It's unclear whether the bomb bounced off the folded roof of the convertible or the archduke himself batted it away. Regardless, the bomb exploded underneath the third car in the procession, injuring several members of the imperial entourage and sending shrapnel flying into the crowd of spectators lining the street.

As chaos broke out, Čabrinović pushed his way through the crowd. He swallowed the cyanide powder as he fled, then jumped over the parapet into the Miljacka River to ensure a swift death. Unfortunately, the cyanide powder was of inferior quality; it seared his throat and stomach lining but didn't kill him. To add insult to injury, the river had largely dried up in the summer heat, leaving Čabrinović vomiting on the sandy riverbank. The failed assassin was soon accosted by a shopkeeper, an armed barber, and two police officers.

As an angry mob descended on Čabrinović, the archduke insisted on stopping the procession so he could check on his friends, who had sustained minor injuries in the explosion. After a short delay, he urged the motorcade forward: "Come on. That fellow is clearly insane; let us proceed with our programme." The Gräf & Stift continued on through the streets of Sarajevo, but the remaining assassins along the parade route lost heart, enabling the motorcade to arrive safely at the city hall minutes later.

A splinter had cut Sophie's cheek, but otherwise the royal couple was unharmed. The mayor, too nervous to improvise, began delivering an ill-timed speech. "All of the citizens of the capital city of Sarajevo find that their souls are filled with happiness," he said to the archduke and his wife, "and they most enthusiastically greet Your Highness's most illustrious visit with the most cordial of welcomes . . ." To this, the archduke exploded with anger, thundering away at the officials there to greet him: "I come here as your guest and you people greet me with bombs!" After a moment, Ferdinand collected his composure and delivered his own speech from prepared notes that were now splattered with the blood of an injured officer from the third car.

After the ceremonial exchanges, the archduke met with officials to discuss his schedule. It was then that Ferdinand decided to skip his afternoon engagements so that he and his wife could go straight to the hospital to visit those who had been wounded in the bombing. When a member of the archduke's staff warned that this could be dangerous, Oskar Potiorek, the governor of Bosnia and Herzegovina, barked, "Do you think Sarajevo is full of assassins?" Everyone's patience was wearing thin.

Along with the governor, the archduke and duchess stepped back into the convertible. Lojka twisted the key in the ignition. In the confusion, nobody notified the drivers of the motorcade that they should take an alternative route to the hospital, and so the cars set off in the same direction that they had come. As a result, the first car turned onto Franz Joseph Street, which was on the original parade route leading to the National Museum that the archduke was scheduled to visit in the afternoon. Lojka followed. It was then that Potiorek realized the error. "This is the wrong way!" he shouted. "We are supposed to take the Appel Quay." Lojka rolled to a stop in order to shift gears. Unbeknownst to him, he had unwittingly presented the archduke as a stationary target to the one man in the crowd who was still determined to kill him.

Gavrilo Princip—who, like Čabrinović, was also dying from tuberculosis and felt he had little to lose—could hardly believe his eyes. He took out his Browning Model 1910 semiautomatic pistol and took aim. Through good marksmanship or just dumb luck, he fatally wounded the royal couple. The first bullet passed through the door of the car, penetrating the duchess's abdomen and rupturing a stomach artery. The second bullet tore through the archduke's neck, severing his jugular vein. As the car sped off, the duchess fell into her husband's lap. Potiorek could hear Ferdinand whispering, "Sophie, Sophie, don't die, stay alive for our children," before he slipped into unconsciousness. Both were dead by eleven o'clock, just hours after they had arrived in Sarajevo.

A crowd descended on Princip, knocking the pistol from his hand as he raised it to his own temple. They kicked and clawed at him and probably would have killed him right then and there had police officers not managed to drag him away. Princip was later tried and sent to prison, where he wasted away from tuberculosis until he weighed less than ninety pounds. He died just weeks before the end of the global war that he had helped initiate.

The assassination was a catalyst of war, setting off a rapid chain of events that destabilized Europe due in part to a web of alliances that bound certain nations together. These alliances meant that if one country was attacked, the allied countries were obligated to defend it. On July 28—one month after the archduke was assassinated—Austro-Hungary declared war on Serbia. The very next day, imperial forces began to shell the Serbian capital of Belgrade. This declaration of war forced Russia to mobilize its troops, since it was bound by a treaty to defend Serbia, which in turn led Germany—allied with Austro-Hungary under the Triple Alliance agreement of 1882—to declare war on Russia. One by one, the fragile bonds of peace holding together the great powers of Europe began to loosen, and nation after nation slid inexorably into what would become the horror of World War I.

The escalating tension on the European continent received limited coverage in the British press. Articles about the situation were often buried deep inside newspapers. A debate about whether boxing was an appropriate spectator sport for women was commanding significantly more public interest. Over two thousand articles appeared on the subject in British newspapers in July 1914 alone, with headlines such as "Women at Boxing Matches. Is Their Presence Unbecoming?" The controversy over the influence of American ragtime music on British youth received similar interest.

The attitude of Britain's politicians toward events on the Continent was likewise dismissive. There was little enthusiasm in Parliament for a war in support of Serbia and her dictatorial ally, Tsarist Russia. Just eleven days before Britain entered the conflict, Prime Minster Herbert Asquith reassured his close friend Venetia Stanley that "happily there seems to be no reason why we should be anything more than spectators." Asquith—whose political party had come to power under the slogan "Peace, Retrenchment and Reform"—was more preoccupied with the looming threat of civil war in Ireland, where the prospect of home rule was dividing Nationalists and Unionists. The gathering storm in Europe seemed far away. By early August, however, it was clear that the coming conflict would not remain just another Balkan quarrel.

On August 3—two days after declaring war on Russia— Germany declared war on her ally France, hoping for a quick victory over the French before the slow-moving Russians could mobilize. Germany immediately began moving troops to the border of Belgium, which had been neutral by treaty since 1839. The German chancellor, however, dismissed the treaty as "a scrap of paper."

Germany forged ahead with its military deployment and offered to pay costs to move its men through neutral Belgium en route to its invasion of France. Germany was convinced that its army would be granted passage, but the Belgians were outraged by Germany's violation of the treaty. Meanwhile, Britain—anxious about the imbalance of power in Europe should Germany conquer France—issued an ultimatum the following day, demanding that Germany withdraw its troops from Belgium. When no response was forthcoming, Britain declared war.

That evening, thousands of people crammed onto the Mall

leading to Buckingham Palace. They waved flags and sang the national anthem. The *Daily Mirror* reported that King George V and his family were "hailed with wild, enthusiastic cheers when they appeared at about eight o'clock last night on the balcony of Buckingham Palace, before which a record crowd had assembled." The mood was jubilant. No one could imagine the reality of the war that was about to unfold. The next day, torrential rain swept the country—a portent of what Britain would face over the next four years.

Newspapers now called for men to step up and do their duty "For king and country." Olive Finch, a Londoner, remembered that "it seemed as though the world had come to an end . . . suddenly there were crowds of men rushing to enlist and hordes of men tramping along the streets in platoons and on top of trams." Sons, brothers, fathers, and husbands from all over Britain abandoned summer holidays and flooded into recruiting depots, promising tearful loved ones that the war would be over soon.

Among these recruits were tens of thousands of underage boys gripped by patriotic fervor and a sense of adventure. One of them was sixteen-year-old Abraham "Aby" Bevistein, who enlisted under a false name and age. His excitement was quickly dampened when he suffered severe shock after a German mine exploded next to him. Frightened and traumatized, Aby fled his post, only to be caught and arrested shortly afterward. He later became one of 306 British soldiers executed for desertion. Their names were sometimes read aloud before the beginning of offensives as a warning to those who were contemplating the same desperate measure. Among the deserters was Private James Smith, who fell to the ground, bleeding but alive, after a firing squad's botched attempt to execute him. His friend Private Richard Blundell shot him in the head at close range after being promised ten days' leave for completing the execution. Seventy-two years later, Blundell lay on his deathbed muttering, "What a way to get leave, what a way to get leave."

Back in Britain, the newly appointed Secretary of State for War, Lord Herbert Kitchener, urged the government to ramp up its recruiting efforts. Kitchener—who had gained notoriety for his "scorched-earth policy" during the Boer War at the turn of the century—foresaw a long, tedious war that spanned years, not months. In a grave speech to members of the Cabinet, Kitchener calculated a three-year conflict that would require the recruitment of a million men or more. Foreign Secretary Edward Grey was astounded. He thought Kitchener's prediction was "unlikely if not incredible" and clung to the idea that the war would be over before a million men could even be trained. Kitchener, however, would not be deterred. Early on, he helped launch an aggressive recruitment campaign to swell the ranks of the regular army. Hundreds of posters bearing a stern image of the Secretary pointing a finger at the viewer with the slogan "[Lord Kitchener] Wants You!" were plastered all over London.

Some young men were spurred to volunteer not by patriotism, but by a fear of being handed a white feather—a symbol of cowardice. Norman Demuth, who was only sixteen years old at the time, remembered someone confronting him one day after he had left school. "I was looking in a shop window and I suddenly felt somebody press something into my hand and I found it was a woman giving me a white feather," he recalled. "For the moment I was so astonished I didn't know what to do about it."

Demuth—who had tried on several occasions to convince the army that he was nineteen—rushed off to the recruiting office with renewed zeal. This time, he was successful. Demuth was eventually wounded and discharged from the army. Before the war was over, another woman pressed a feather into his palm while he was riding a bus. "Oh Lord, here we go again," he thought to himself. He proceeded to use the feather to clean out his pipe before handing it back to her and remarking, "You know, we didn't get these in the trenches."

In the end, Kitchener's recruitment campaign was a roaring success. Over half a million men enlisted within the first two months of the war. By the end of 1915, over three million soldiers were serving in the British armed forces. The Secretary of State for War had succeeded in producing the greatest volunteer army Britain had ever seen.

As the number of enlistees grew, so too did the need for medical workers to care for the war's sick and wounded. The Royal Army Medical Corps operated the army's medical units, bolstered by voluntary help from such bodies as the British Red Cross, St. John's Ambulance, and the Friends' Ambulance Unit. For those wishing to serve as medics, there were many organizations to which they could apply.

Civilian women turned out in the thousands to volunteer as nurses. Many came from the middle and upper classes and had never set foot inside a hospital. On the wards, they were called upon to perform tasks requiring domestic skills few of them possessed. "I can see a girl now sitting on the stairs with a duster, wondering what on earth to do with it," one woman recalled. Romantic notions of nursing were soon shattered by the grim reality of bedpans, vomit, and blood. Young women who had never even seen a man in his underclothes were suddenly expected to work with the mutilated bodies of soldiers evacuated straight from the trenches.

Enid Bagnold, a British playwright who volunteered at the start of the war, remembered severed legs piled high in baskets outside the door to the operating room: "[t]he wounded came just as they were, their bandages soaked in blood . . . Operations went on without stopping." Women from all over Britain suddenly found themselves in similarly traumatic situations. Claire Elise Tisdall, a volunteer nurse working in London, watched as a soldier was stretchered past her one night. In the dim light, she thought the lower half of his face was covered by a black cloth. Only later did she realize it had been completely blown off.

Not all the women who volunteered lacked formal training. When war broke out, qualified nurses saw an opportunity to put their professional skills to use. "It is at a time like this that a trained nurse proves her worth," one recalled. "It is impossible for the surgeons to see all the patients as they come in, so the Sisters and nurses must do the best they can." There was often tension between those who were qualified and those who had little or no experience caring for the sick and wounded. One professional nurse complained that the skills acquired through formal training could not "be imparted in a few bandaging classes or instructions in First Aid." Such conflicts notwithstanding, both inexperienced and experienced nurses were eagerly accepted into service.

Female doctors, however, faced more stubborn obstacles when it came to finding outlets for their hard-earned skills. When Dr. Helena Wright tried to secure a post in a military hospital, she repeatedly encountered sexist resistance. Elsie Inglis a respected physician and suffragist—faced similar prejudices. When she wrote to the British War Office to suggest that female medical units be allowed to serve at the front, she received the reply, "My good lady, go home and sit still." This did not deter Inglis, who eventually offered her services to the French and would go on to set up women's units in not only France but also Serbia, Corsica, Greece, Malta, and Russia.

At this early stage of the war effort, the Royal Army Medical Corps, along with a multitude of voluntary medical organizations, were simply overwhelmed by the deluge of people clamoring to lend a hand. Among the crush of medical volunteers was thirty-two-year-old Harold Gillies, who had signed up with the Red

Cross shortly after Britain joined the conflict. In January 1915, he was finally called up. Gillies took a leave from Dr. Rees's office and packed his bags for France.

The decision to volunteer could not have been an easy one, as he had to leave behind his pregnant wife, Kathleen, who would give birth to their second child, Margaret, a few weeks after his departure. Being separated from his growing family was hard enough. Soon, Gillies would also discover that the medical crisis on the Western Front was a far cry from extracting scissors from the rumps of Covent Garden's ballerinas.

❖ 2 ❖

THE SILVER GHOST

The Rolls-Royce cruised at a stately pace through the narrow streets. Nicknamed the "Silver Ghost" due to the wraithlike quietness of its engine, the open-topped car with a cream exterior and rust-colored leather seats had become a familiar sight to army personnel occupying the coastal city of Boulogne in northern France. At the wheel was a sandy-haired man with a fleshy face named Auguste Charles Valadier. He was a Franco-American dentist who was famous among the troops for having converted his luxury car into a mobile operating room by retrofitting it with a dental chair, drills, and equipment—all at his own expense. It was this eccentric figure, with his highly polished riding boots and glittering spurs, who would reveal to Harold Gillies the desperate need for reconstructive surgeons near the front.

Valadier was born in Paris on November 26, 1873, to Charles Jean-Baptiste and Marie Antoinette Valadier. When he was a child, he sailed to the United States with his father, a pharmacist, and two younger brothers. As the ship slowly lumbered into New York's harbor, he may have caught a glimpse of the pedestal being constructed on Bedloe's Island that would soon support the Statue of Liberty—an extraordinary gift from his native country to his adoptive home.

Before Valadier was allowed to disembark, he was first examined by doctors who boarded the ship to inspect passengers for communicable diseases, such as smallpox, yellow fever, and cholera. Those who were thought to be infectious were separated from other arrivals and placed in quarantine for an indefinite period. After receiving a clean bill of health, Valadier and his family were ushered into a cylindrical sandstone fort off the tip of Manhattan known as Castle Clinton. Before the establishment of Ellis Island in 1890, Castle Clinton was where foreigners entering the United States via New York were processed. It's unlikely that any of the Valadiers would have been carrying paperwork with them when they arrived. All that was required at the time was a verbal confirmation of one's name and country of birth.

Valadier spent most of his childhood in the United States and eventually became a naturalized citizen like his father. When the time came to choose a career, he set his sights on dentistry, enrolling in the Philadelphia Dental College—the second-oldest dental school in the country. After his father died, Valadier moved to New York City, where he opened a dental practice at 39 West Thirty-Sixth Street, not far from where the Empire State Building would be erected several decades later. Valadier might have remained in New York his entire life had the sudden death of one of his brothers not prompted him to return to Paris in 1910. This relocation was due in part to Valadier's mother, who, having recently been widowed, was now rumored to be the mistress of the famed newspaper publisher James Gordon Bennett, Jr. As such, she was flush with cash, and she lured her son back to Paris with the promise of a lavish fifth-floor apartment on the fashionable Place Vendôme.

On his return, Valadier studied at l'École Odontotechnique

de Paris in order to obtain his dental certification in France. Afterward, he set up a practice on Avenue Hoche, a stone's throw from the Arc de Triomphe in an affluent neighborhood near the Champs-Élysées. There he treated several noteworthy clients, including the king of Spain. Valadier was always at the forefront of his profession. At the Pasteur Institute, he spent considerable time collaborating with a doctor named H. Spencer Brown on the use of certain vaccines for the prevention of severe gum disease. He was just as interested in preventive care as he was in curative measures.

Valadier could be quick to anger. In 1913, a year before the war broke out, he married his second wife, Alice Wright, granddaughter of the former United States Minister to Brazil. Halfway through the ceremony, a church official whispered something in the ear of the presiding priest, after which the ceremony slowed perceptibly. Later, Valadier could be heard thundering away in the vestry, "The bastard! They made me pay 25,000 francs!" It turned out that the priest had been told that Valadier was a divorcé, which made him ineligible to get married in the Roman Catholic Church.

Valadier quickly settled into domestic life, moving into an even grander apartment at 47 Avenue Hoche, close to his dental practice. Then, on August 1, 1914—while he was attending the annual meeting of the American Dental Society of Europe in Paris—France declared war on Germany. The world was suddenly thrown into turmoil.

Given his medical background and location, Valadier seemed an obvious candidate for service with the French army. However, rules stipulated that he would either have to enlist as a private or volunteer with the Foreign Legion due to his U.S. citizenship. Neither of these options appealed to the successful forty-year-old professional who had spent much of his life hobnobbing with the upper echelons of society. Moreover, the French army did not have an organized dental corps at that time, and dental care was provided to soldiers in a haphazard manner that Valadier would likely have detested.

Unfortunately, the British didn't value the skills of dentists any more than the French did. Ever since the invention of the repeating rifle in the mid-nineteenth century, the conservation of teeth was a low priority for the army, as infantrymen were no longer required to bite open paper gunpowder cartridges as part of the loading procedure. The military's failure to heed the adage "an army that cannot bite, cannot fight" had been causing problems for some time. During the Boer War, 6,942 men were admitted to hospitals due to dental issues: one third had to be invalided back to Britain, while the others remained unfit for full active duty. Despite these hard lessons, the army was slow to act.

At the start of World War I, general army surgeons were expected to tend to the dental needs of soldiers. Consequently, not a single dentist accompanied the British Expeditionary Force to France in August 1914. Henry Percy Pickerill, a dental surgeon who would later play an important role in the development of plastic surgery, bemoaned the fact that general surgeons were initially in charge of the oral healthcare of the troops. "Can a medical man say just exactly from the necessarily hurried examination he must give of the mouth, and without a special dental knowledge, what constitutes a good dentition?" he asked shortly after the war began.

The decision not to send dentists with the British Expeditionary Force seems odd, given the poor state of working-class teeth at the time. After interviewing several soldiers in his company, one officer recorded "numbers of men frankly stating that they had never used a tooth brush in their lives." Additionally, basic army food presented challenges to already fragile teeth. High-calorie biscuits produced under government contract by Huntley & Palmers were notoriously hard and could crack a soldier's incisors if not first soaked in tea or water.

The monotonous diet and lack of oral hygiene also led to the painful condition of acute necrotizing ulcerative gingivitis, commonly known as "trench mouth." It occurs when bacteria build up in the mouth, causing bleeding, ulcers, and bad breath. In its advanced state, it can also cause the gum membranes to slough off. Left untreated, a man might experience difficulties eating and swallowing. It was a vicious cycle that took a terrible toll on the health of many soldiers during the war.

Defective teeth were not just a problem for the army once men were ensconced in the trenches. They were also a major reason for rejecting recruits. Jokes about the state of the nation's teeth even reached the pages of the satirical British weekly magazine *Punch*. Shortly after the war broke out, the magazine published a cartoon that depicted an incredulous man at a recruiting office protesting the decision to turn him away because of his rotten teeth: "Man, ye're making a gran' mistake. I'm no wanting to bite the Germans, I'm wanting to shoot 'em." It turns out that the cartoon was no fiction. It reflected the experiences of real civilians. The British journalist and author Robert Roberts recalled in his memoir a conversation between his mother and a Boer War veteran named Mr. Bickham, who had been rejected by the army on account of his bad teeth. "They must want blokes to bite the damned Germans," he shouted at the recruitment officers in frustration. As he walked away, he called over his shoulder, "They'll be pulling me in, though . . . before this lot's done!"

At the beginning of the war, dentists acted unofficially to ready infantrymen for battle by offering them free dental services before they were sent abroad. This ensured that the greatest number of recruits could make it to the front. C. V. Walker, a dental student at the Newcastle-upon-Tyne Dental Hospital, remembered extracting over *nine hundred* teeth in a single month. The hospital was overwhelmed by patients, so Walker was limited to "two syringefuls [sic] of a solution of cocaine [per man]"—cocaine being a commonly used dental anesthetic at the time. One soldier, who had all his upper teeth extracted, asked Walker to remove the lower teeth as well. When Walker explained that he would have to do it without anesthetic, the man replied enthusiastically, "I'll have them out with nowt as I want to get out to the front."

And yet, despite the prevalence of dental decay, it was only when General Douglas Haig developed an excruciating toothache at the height of the Battle of Aisne in October 1914 that the absence of dentists *within* the army was finally addressed. Valadier, whose professional reputation preceded him, was summoned from Paris. As he traveled the seventy-odd miles to Aisne, he was forced to abandon his vehicle and proceed on foot when the car was struck by German shells. According to one account, once he reached the general's headquarters, he extracted the rotten tooth "under a rain of bullets." (Haig would later recommend that Valadier be decorated for his "excellent and most valuable surgical work on the jaw performed gratuitously by this gentleman for all ranks of the British Army.")

So it came about that Valadier was given a temporary commission in the Royal Army Medical Corps. Shortly after General Haig's successful extraction, Valadier received the commission and the honorary rank of "Local Lieutenant" in the British Army. He was one of the first dental surgeons to provide treatment to British troops in France during World War I. By the end of 1914, there were twenty dentists acting in an official capacity. This number gradually grew, until there were 831 dentists serving at the time of the Armistice in 1918. It wasn't until after the war, in 1921, that a dedicated Army Dental Corps would finally be established by the Secretary of State for War, Winston Churchill.

But in this First World War, the need for dentistry skills that far exceeded the ability to pull a rotten molar was becoming acute, as Valadier would soon find out. In October 1914, he was assigned to No. 13 Stationary Hospital, which had been set up in abandoned sugar sheds at the Gare Maritime in Boulogne just weeks earlier.

Valadier's arrival was timely, as the First Battle of Ypres had just begun, and his services would soon be in high demand.

Early in the war, the tremendous number of casualties being generated on the Western Front necessitated the creation of base hospitals for the triaging and evacuation of the sick and wounded. These were built close to railway lines and near ports to expedite the reception of continuous streams of injured men and as staging posts for the evacuation to Britain of those who needed long-term treatment. Base hospitals were further broken down into two categories: general or stationary. In theory, the former were larger, with a capacity of one thousand patients or more, while the latter were smaller and more specialized. No. 13 Stationary Hospital, to which Valadier was assigned, was one of the first of its kind. And it wasn't long before patients began to overwhelm medical personnel.

In October 1914, No. 13 Stationary Hospital became a veritable warehouse of human suffering during the First Battle of Ypres. The fighting was concentrated around the ancient Flemish city of Ypres, whose fortifications guarded routes to the English Channel and access to the North Sea beyond. The battle raged for a full month, with both sides sustaining enormous casualties. One of Valadier's colleagues recalled the chaos that greeted her upon her arrival at the hospital: "the sheds were being converted into wards; wooden partitions were being run up, bedsteads carried in, the wounded meanwhile lying about on straw or stretchers." As thousands of injured soldiers crowded into the former sugar sheds that now formed No. 13 Stationary Hospital, Valadier realized there were far graver challenges than the ubiquitous tooth decay awaiting him at the front.

Trench warfare resulted in a high number of survivable neck and head wounds, but the waterlogged and richly manured soil of the local area caused a spike in infection rates. This was especially true in soldiers who had been hit by high-explosive shells that caused severe lacerations to the skin and underlying tissue and carried bacteria deep inside the body. Mortality rates from wound infections reached as high as 28 percent in some base hospitals. And this did not account for those who died en route or at the front itself.

Ironically, the surgeon Joseph Lister—who is credited with saving tens of thousands of lives in the nineteenth century by introducing antiseptic techniques to the practice of surgery—was indirectly responsible for the high incidence of sepsis in Europe at the start of the First World War. His success meant that the latest generation of surgeons, brought up on germ theory and the principles of asepsis, were unaccustomed to identifying and treating infected wounds, since they rarely encountered them in their day-to-day practice. But the rich farming soil of France and Belgium harbored lethal microbes that caused tetanus, gas-gangrene, and septicemia. The battlefield was a breeding ground for pathogens, so much so that the Australian medical officer Arthur Graham Butler called it "a war of faecal infection—streptococcus and anaerobes." A surgeon quite literally sealed a soldier's fate when he sutured a wound packed with bacteria. Even if antiseptic dressings were applied afterward, they did little to address deep-seated infections.

Fortunately, Valadier recognized that bad situations were made worse by the premature closure of facial wounds before they could be properly cleaned out—especially as this area of the body was already teeming with bacteria from poor dental hygiene. Early on, he began thoroughly irrigating facial wounds before treating them. To that end, he devised a mobile apparatus, which he called the "fire engine." It consisted of a large drum containing boiled water, with rubber tubing fitted to it. To provide pressure, a bicycle pump was connected to the drum.

It wasn't long after arriving in Boulogne that Valadier convinced the general staff at the hospital to allow him to establish a temporary jaw unit in nearby Wimereux to handle the large number of maxillofacial injuries arriving from the front. He paid for most of the medical equipment out of the profits of his private practice back in Paris, as well as from money he had inherited after his mother's recent death. (He also offered his services entirely gratis until October 1918—one month before the war ended—when he finally accepted the pay and allowances of a Major of Infantry.)

It was at this specialist unit that Harold Gillies first met the dentist, and his own career took an unexpected turn. He had been appointed to supervise Valadier's work, since the Frenchman's dental credentials didn't allow him to operate without oversight from a surgeon. As an ENT specialist with a deep understanding of head and neck anatomy, Gillies was uniquely qualified for the job. But it was arguably Gillies who benefited most from this early partnership. Not only did it teach him the value of dentistry to the practice of facial reconstruction, it also demonstrated to him the transformative power of plastic surgery.

The term "plastic surgery" was coined in 1798 by the French surgeon Pierre-Joseph Desault. Before the manufacture of the synthetic material known today as plastic, the word often referred to an object that could be shaped or sculpted—in this case, a person's skin or soft tissue.

When Gillies met Valadier in 1915, plastic surgery was still in its infancy as a medical specialty. Most surgeons had little, if any, experience in dealing with the widespread destruction of soft tissues of the face. Efforts in earlier periods to rebuild, repair, or alter the appearance of the face were typically confined to small areas, such as the nose or the ears. But even the most basic operations were not widely practiced despite the fact that developments in anesthetics in the latter half of the nineteenth century had made them less painful than in previous periods. Both reconstructive and cosmetic procedures remained rare prior to World War I. Going under the knife for an experimental form of surgery posed serious risks in terms of infection and could lead to further disfigurement if the surgery was done incorrectly.

It wasn't until the American Civil War that systematic attempts were made to reconstruct large sections of faces. This was driven in part by the horrific damage caused by a new type of ammunition: the conical bullet. Known as a "minié ball," it was a projectile that flattened and deformed upon impact, causing maximum destruction.

Arguably the most skillful and imaginative surgeon to emerge from the Civil War battlefield was Gurdon Buck, one of the founders of the New York Academy of Medicine and one of the very first doctors to include pre- and post-operative photos in his publications. In one noteworthy case, Buck reconstructed the lower jaw of William Simmons, a soldier who had been hit in the face by an exploding shell. Despite the severity of the wound, Buck was able to restore some function to the injured area. Before long, Simmons— who was unable to eat or speak when he was first wounded—no longer showed any reluctance to engage in conversation.

One of Buck's most complicated cases was not the result of Civil War weaponry, but rather a side effect of medication administered to a soldier suffering from typhoid fever. Shortly after being given calomel (mercury chloride), Private Carleton Burgan developed an ulcer on the tip of his tongue. This soon became gangrenous, spreading to other areas of his face. Within weeks, gangrene had eaten away his palate, right cheek, and right eye.

Buck enlisted the help of the dentist Thomas B. Gunning, who created a plate made of hard rubber as a substitute for the missing palate, with another piece above it that filled out the right side of the nose. Buck then performed a series of operations, including a rudimentary skin graft, that helped restore Burgan's appearance. The wounded soldier wore the artificial palate for the rest of his life. He went on to marry and have eight children, dying at the age of seventy-two.

Although there were surgeons who attempted to restore function to damaged faces during the Civil War, few paid attention to aesthetics. When Private Joseph Harvey was hit by a shell fragment at the Battle of Chancellorsville, it ripped through his right cheek, chipping off part of his lower jaw and destroying his eye. He was admitted to Mansion House Hospital in Alexandria, where a surgeon exfoliated jagged portions of the bone. Harvey was then released, despite the fact that a giant hole remained in his cheek, through which saliva and other liquids leaked. He eventually found employment as a night watchman and died a few years later, likely due to complications from his injury. Like Private Harvey, most soldiers who sustained head injuries at this time were left with terrible wounds and gaping holes in their faces.

Plastic surgery remained the exception rather than the rule during the Civil War. Gurdon Buck was one of only a handful of surgeons willing to attempt such risky procedures. As a result, fewer than forty "plastic operations" are reported to have taken place in both the North and the South. However prevalent facial injuries were in the 1860s, they were far more pervasive during the First World War. Because of this, the surgical inadequacies of past centuries would finally be addressed, paving the way for plastic surgery to enter a new era—one in which methods could be tried and tested on a scale hitherto unimaginable.

At the jaw unit in Boulogne, Gillies soon learned that Valadier was a divisive character, either loved or hated by those who knew him. One colleague described him as "a charming, jaunty cowboy" who could roll cigarettes with one hand while holding the reins of a horse with the other. Ferdinand Brigham—an American surgeon who worked with Valadier during the war—had a slightly different opinion of the dentist, whom he called "an impresario, a pretender, a ham." When Valadier presented Brigham with a handsome little edition of his own poems, the American surgeon wondered aloud who had written them on the Frenchman's behalf.

Although some of his peers may have held him in contempt, Valadier was revered by the troops for not accepting payment from the army for his services, which led to great speculation about his alleged wealth. Ellis Williams was a Welsh infantryman who was treated by Valadier after being injured in the Battle of Mametz Wood. He remembered the rumors surrounding the unusually tall Frenchman. "It was said that he was a millionaire," Williams recalled. "Judging from his attire I would not be surprised. He would buy the best clothes . . . He was a very smart-looking man."

Gillies was more interested in Valadier's work than in his wardrobe and witnessed his early experiments with bone grafts— effectively the transplantation of pieces of bone. Grafts were nothing new in the history of medicine. In 1668, the Dutch surgeon Job van Meekeren became the first to describe the procedure when he documented the case of a Russian surgeon who had repaired a defect in a soldier's cranium by implanting a fragment of dog bone into the man's skull. The Church deemed the surgical practice "unchristian" and excommunicated the soldier as a result. When the man requested that the procedure be reversed, his surgeon discovered that his natural bone had grown around the graft. As a consequence, it could not be removed.

Despite the success of some of these early cases, most surgeons in the seventeenth and eighteenth centuries were hesitant to perform such an invasive procedure. It wasn't until the discovery of anesthesia and the development of antisepsis in the nineteenth century that bone grafts became more widespread. During this time surgeons began experimenting with autografting, which involved removing bone from one part of the body of a patient and transferring it to another.

Working alongside Valadier, Gillies realized his surgical skills would be for naught unless he figured out a way to reliably reconstruct faces. Without the use of bone grafts, attempts to close facial wounds often resulted in the severe distortion of the features, which was not only considered unsightly, but could also interfere with a patient's ability to speak or eat. In one case, Valadier replaced two-and-a-half inches of bone in a man's shattered jaw. In another, he took a section of a patient's rib and inserted it beneath the flesh of his forehead in order to reconstruct the soldier's nose. Once the graft took hold, Valadier reshaped the skin around it. Gillies would replicate and improve upon these types of procedures later in the war.

In one of his more harrowing cases, Valadier helped rebuild a man's jaw after it had been blown off his face completely, leaving only the soldier's "gullet" exposed. Ellis Williams, the Welsh soldier who had passed comment on Valadier's impeccable wardrobe, watched as his friend Jock underwent several excruciating procedures:

> Firstly, he carved the shape of the jawbone in silver and adapted it so it worked in the front of the jaw. He then cut a piece of skin on [Jock's] chest, turned it over and sewed and shaped it alternatively until it became the right shape. He then placed false teeth on it and when he had finished, Jock could eat everything and you couldn't say there was anything wrong with him.

Time and again, Valadier proved to be an innovative and resourceful practitioner.

In two separate cases involving fractured mandibles, he performed what is now known as distraction osteogenesis: a technique for creating bone without the need for grafting. Although the practice dates back to the sixteenth century, when Teutonic knights used leg braces and screw mechanisms to straighten limbs, it had never been performed on the face until Private Philip Thorpe of the King's Liverpool Regiment was hit by a shell, severing most of his lower lip and a large portion of his jaw.

Valadier wired the two ends of Thorpe's mandible together and attached an expansion screw to a vulcanite (rubber) plate that was then used to slowly push apart the fractured ends. This helped stimulate new bone growth, which formed to fill in the gaps caused by the stretching of the mandible. Thorpe recalled the moment Valadier examined the X-ray of his jaw to see what progress, if any, had been made: "He suddenly put down the plate, grabbed the ward sister round the waist and pranced up and down the ward shouting 'We've done it, we've done it.'" Distraction osteogenesis only became a mainstay of maxillofacial surgery in the early 1990s, which illustrates just how far ahead of his time Valadier was.

Gillies was influenced and inspired by Valadier during those months he spent in France. He later praised the dentist for his pioneering role in facial reconstruction: "[t]he credit for establishing the first Plastic and Jaw Unit, which so facilitated the later progress of plastic surgery, must go to the remarkable linguistic talents of the smooth and genial [Valadier]." It was the eccentric dentist with his strangely outfitted Rolls-Royce who had convinced generals strapped into his dental chair of the need for a specialist unit.

The practices of dentistry and medicine formed an interdisciplinary two-way street. Dentists like Valadier learned much from their surgical colleagues during the war. "It was therefore inevitable," Gillies later wrote, "that the era of teamwork in surgery should begin—dental and plastic teams joined forces." Long after Gillies left the front, he and Valadier would continue to correspond. Occasionally, Valadier would even refer patients to Gillies, when the limitations of his own skills prevented him from helping some of the more severe cases.

While working with Valadier, Gillies encountered patient after

patient suffering from severe head and neck wounds. It soon became evident to him that these casualties would be best served
by the establishment of a permanent specialist unit in Britain, to
which all those requiring facial reconstruction could be sent. He
later recalled: "I felt that I had not done much to help the wounded
and that I must bestir myself." A seed had been planted in Gillies's
mind—and as he took up his next assignment, it would be nurtured in the brutal hothouse of frontline surgery.

⇥ 3 ⇤

SPECIAL DUTY

The Belgian Field Hospital and its staff had already endured a good deal of turmoil by the time Harold Gillies arrived in the spring of 1915. It was the hospital's third location since the war had started. Its first incarnation was established in Antwerp, until German troops besieged the city in October 1914. Under cover of darkness, doctors and nurses narrowly escaped enemy capture by loading patients quickly onto motor buses already crammed full of crockery, blankets, and medical instruments. Although the buses became separated in the chaos of the bombardment, the staff eventually regrouped and established a second hospital in a converted schoolhouse in Veurne, fifteen miles east of Dunkirk. But this sleepy Belgian town would not escape German attention for long.

In January 1915—three months after the Siege of Antwerp—Veurne was heavily shelled. As hospital staff once again began evacuating, a thirty-three-year-old nurse named Rosa Vecht was hit in the leg with shrapnel while saying goodbye to colleagues. The leg was so badly damaged that it had to be amputated at the hip joint. During the operation, however, Vecht lost a tremendous amount of blood and succumbed to her injuries. She was the only Dutch nurse to die during the First World War.

The third location of the Belgian Field Hospital was just two miles from the second, in a provincial village called Hoogstade, which was accessible only by narrow, slippery roads. Staff converted a dilapidated two-story almshouse into an eighty-bed hospital, behind which were farm buildings nestling in fields that sloped toward a babbling brook. Nurses, orderlies, and laundry maids slept in the attic, where they tied bandages to the rafters and pinned sheets to them in order to create partitions. On the ground floor was the kitchen, headed by a chef who had escaped German captivity by cooking a large banquet for his captors, lacing the dishes with liqueurs and plying attendees with gallons of wine. After he had gotten the enemy suitably drunk, he grabbed his hat and coat and calmly walked away to freedom.

Like its predecessors, the Belgian Field Hospital at Hoogstade was uncomfortably close to the action. *The Times* declared it "The Nearest Hospital to the Firing Line" and encouraged readers to help save lives and limbs by donating to the cost of its upkeep. And yet the atmosphere surrounding the hospital was an odd mix of the pastoral and the martial. One nurse recalled:

> Cows roamed around amongst motor ambulances and cars painted war-grey with huge red crosses upon them. Soldiers carried in stretchers of wounded or carried out the dead, nuns sat milking cows, infirm old almshouse women in large mob caps pottered about, while army nurses flew past on divers errands. Mechanics mended car machinery and rough ploughmen beat the corn with old-fashioned flails in the same barn.

The rural setting might have had charm, but it lacked access to all the modern conveniences that a city might offer. The hospital was not just surrounded by farmland. It was also surrounded by mud and muck. "A great cesspool ran under a large part of the farm yard quite close to the well—where the pump-handle squeaked day and night—from which we got all the drinking water," the same nurse wrote in her diary. Suffice to say, newcomers to this outpost did not experience it as a home away from home.

On May 1—just one week before the infamous incident in which the RMS *Lusitania* was torpedoed by a German U-boat, drowning 1,201 civilians, including 128 Americans—Gillies landed at Dunkirk. With him was Herbert W. Morrison, the newly appointed commandant of the Belgian Field Hospital. Unfortunately, the hospital staff hadn't been notified of their arrival date, and so, when Gillies and Morrison stepped onto the quay, no one was there to greet them. The pair waited several hours until a driver named Nat Batten finally arrived to pick them up. By then, Morrison was positively furious. He complained about the apparent lack of organization at the hospital and blamed Batten personally for the oversight. Batten did not take this outburst lightly and warned Morrison that he would not hold his position as commandant long with that attitude. As it turned out, Morrison would later have Batten removed—initiating a chain reaction of resignations in the wake of this unpopular act.

Gillies and Morrison weren't the only new arrivals that spring. The illustrious scientist Marie Curie also visited the hospital. By then, Curie was famous for her discovery of radium. In 1903, she became the first woman to receive the Nobel Prize, and she went on to win it again in 1911. When the war broke out, Curie set aside her scientific pursuits, sealing her entire stock of the radioactive element in a lead-lined container and transporting it to a safe-deposit box in Bordeaux to prevent it from falling into the hands of the Germans. She then redirected her talents to the war effort and created a vehicle that contained a hospital bed, a generator, an X-ray machine, and photographic darkroom equipment. These "petite Curies," as they became known, could be driven right up to the battlefield. The world-renowned physicist and chemist also set up 200 stationary X-ray clinics during the war and helped train 150 women as radiology technicians to help run them.

Curie drove to the Belgian Field Hospital with her daughter Irène in one of her specially equipped ambulances. A nurse there recalled how the scientist would rise each morning at five o'clock before beginning work. "For two or three weeks she lived with us, sharing our daily life, sitting next to us at meals, the most unassuming and gentlest of women," the nurse wrote in her diary. Irène—whose mother taught her to operate the X-ray machine—remained at the hospital long after Curie departed.

The war was prompting eminent personalities like Curie to play their part, but Gillies had yet to make his mark in his chosen field. It was the early days of May 1915, and the Second Battle of Ypres was in full swing when Gillies first stepped into the field hospital's makeshift operating theater. The ramshackle building shuddered under the incessant discharge of guns in the distance. New casualties requiring Gillies's attention arrived at the hospital on an almost hourly basis. Surgeries were performed day and night, with two operating tables occupied at all times. The floor was slick with blood. One nurse recalled how even the ambulance drivers delivering new casualties would "seize a mop and pail and swill up some of the blood from the sloppy floor, or even hold a leg or arm while it was sawn off." The work was endless and grim.

For the newly minted surgeon-in-chief Gillies, the Second Battle of Ypres was a baptism by fire. It not only claimed tens of thousands of lives, including that of fourteen-year-old John Condon—long believed to be the youngest soldier to die in the war—but it was also the first battle in which chlorine gas was successfully employed as a chemical weapon by the Germans.

Chlorine is a diatomic gas—about two and a half times denser than air—that reacts with water in the lungs to form hydrochloric acid, which can cause permanent tissue damage, leading to death in a very short span of time. Lendon Payne, a British soldier, recalled the shocking effects of a gas attack shortly before Gillies arrived in Belgium. "I could hardly believe my eyes when I looked along the bank," he wrote. "[It] was absolutely covered with bodies of gassed men. Must have been over a thousand of them. And down in the stream, a little bit further along the canal bank, the stream there was also full of bodies as well."

Chlorine's usefulness was short-lived, since its color and odor made it easy to detect. It was eventually replaced with deadlier and more effective chemicals such as phosgene, bromine, and bis(2-chlorocthyl) sulfide known colloquially as mustard gas. The use of chemical weapons eventually spurred the development of the eerie gas masks that became iconic of the First World War. But back in the spring of 1915, there was no protective gear, nor was there anyone working at the Belgian Field Hospital who had ever witnessed the catastrophic effects of poison gas on men. One nurse recorded how helpless she felt when faced with casualties of chemical warfare: "There they lay, fully sensible, choking, suffocating, dying in horrible agonies. We did what we could, but the best treatment for such cases had yet to be discovered, and we felt almost powerless." Those who succumbed to the gas were placed in ambulances and driven up the road to their final resting place in a nearby cornfield.

Gillies would certainly have felt the psychological and physical strain while working at the Belgian Field Hospital. A month of fighting left tens of thousands dead and wounded on both sides. By the time he left, he had gained considerable experience in dealing with the effects of military hardware on the human body. But though he could in many cases stop the bleeding and stabilize his patients, he had not yet figured out how to address the long-term damage being wrought upon those at the front.

• • •

In June 1915—not long after the Second Battle of Ypres had ended, resulting in significant gains by the Germans—Gillies was given the rank of major and reassigned to the Allied Forces Base Hospital in Étaples, France. Before taking up his new assignment, he returned to London on leave. It was a welcome opportunity to visit Kathleen, who had given birth to their second child, Margaret, in his absence. As busy as domestic life must have been with two young children, Gillies couldn't resist sneaking in a few rounds of golf. The sportswriter Henry Leach reported that a number of people were "prepared to swear that recently they saw [Gillies] . . . in the uniform of a Major of the R.A.M.C." Leach went on to write that "this Major has once or twice appeared on a certain excellent London course and made most phenomenally low scores upon it."

Gillies didn't spend his entire leave engaged in frivolous activities. He made it his business to investigate the work of others making strides in the branch of surgery that had captured his attention while he was working with Valadier. A colleague lent him a German book that contained a section on jaw surgery. The German medical community had stolen a march on other nations due to recent conflicts like the Balkan War of 1913, which caused numerous facial injuries similar to those that would be seen in the First World War, albeit on a much smaller scale. The Germans were quick to establish a multidisciplinary approach, involving not only surgeons but also dentists and dental technicians to manage various aspects of reconstruction. Most of the operations they performed were unsuccessful, but they had gained much knowledge. The book made a deep impression on Gillies, who remarked, "I had struck a branch of surgery that was of intense interest to me."

This new fascination prompted further investigation. On returning to France in June, Gillies stopped in Paris to seek out a key source of information: Hippolyte Morestin. This surgeon had famously treated the actress Sarah Bernhardt after she shattered her knee jumping from the parapet of the Castel Sant'Angelo in the final act of Victorien Sardou's *La Tosca*. Gillies had heard that Morestin was performing miraculous feats at the Val-de-Grâce hospital, where French soldiers who had sustained facial injuries were being treated. He was eager to observe firsthand Europe's premier facial surgeon at work.

Morestin, a reclusive man with piercing eyes and a pointed goatee, was prone to extreme mood swings. The French writer Georges Duhamel, who undertook surgical training with Morestin before and during the war, described how in the space of a moment, the brooding surgeon could transform into a "wild beast, swift and ferocious." He recalled a time when Morestin rapidly cut out the cancerous tongue of a patient who was only half-unconscious while he "coughed [and] spat blood in our faces." On another occasion, the surgeon heaped recriminations on a dead girl, "speaking to her and overwhelming her with reproaches" for dying before he had a chance to operate. Even in his obituaries, he was remembered as having an uneven temperament, with periods of quiet punctuated by frenzied activity: "Few surgeons have ever shown a more powerful energy than this man with a frail body, an emaciated face, [and] fiery eyes."

Morestin's cantankerous disposition was perhaps understandable, given his past. He was no stranger to trauma himself. In 1902, while he was studying medicine in Paris, a volcano that had lain dormant for decades on the island of Martinique suddenly erupted and spewed hot ash and dust over his hometown of Saint-Pierre. Within sixty seconds, the entire area was razed by pyroclastic material with a temperature of nearly two thousand degrees Fahrenheit. All but two of Saint-Pierre's thirty thousand inhabitants were killed, including twenty-two members of Morestin's own family. Among the dead was his beloved father, also a surgeon, whose encouragement to pursue a career in medicine had saved his son's life by taking him far from the scene of disaster. Since then, Morestin had carried a deep sadness within himself.

Nevertheless, he did not let personal tragedy derail his studies. He exhibited a remarkable understanding of anatomy, which served him well early in his career, when he specialized in the removal of cancerous growths from patients' cheeks and mouths. It wasn't long before he turned his attention to reconstructive techniques to address the scars left behind by his knife.

In contrast to many of his contemporaries, Morestin paid special attention to questions of aesthetics, even going so far as to experiment with mini-facelifts, using a series of small elliptical excisions to tighten the skin. When the First World War broke out, it was only natural that he should focus on maxillofacial injuries. But in his arrogance he believed that *only* a surgeon could repair a face. For this reason, Morestin rarely employed dentists, whom he treated with condescension. Nonetheless, his peerless surgical skills were in great demand. The influx of injured French soldiers increased until he suddenly found himself responsible for no fewer than 480 beds.

The Val-de-Grâce already had a long history of treating military personnel by the time the first casualties began arriving from the front. Rising high above the crowded streets in the heart of what is now Paris's 5th arrondissement, the ornately domed building looked like something from a fairy tale, and for good reason. It was originally built as a church in the seventeenth century to celebrate the miraculous birth of Louis XIV, after his mother, Queen Anne of Austria, had gone childless for twenty-three years. During the French Revolution, Benedictine nuns at the church provided medical care for injured revolutionaries, thus sparing the Val-de-Grâce the desecration and vandalism that so many Parisian churches, such as Notre-Dame, had suffered. Shortly afterward, the building was converted into a military hospital. When Gillies arrived in the summer of 1915, it was a hive of activity. Hippolyte Morestin made an immediate impression on Gillies with his "dagger-like sharpness . . . pointed moustache and tapering beard."

The elder surgeon welcomed Gillies into his operating theater. There, Gillies observed Morestin meticulously remove a cancerous growth from a patient's face and close the resultant wound using a large flap of skin from the subject's neck. Gillies later recalled this turning point in his career: "I stood spellbound as he removed half a face distorted with a horrible cancer and then deftly turned a neck flap to restore not only the cheek but the side of the nose and lip, in one shot." Years later, after he had established himself as a leading authority in facial reconstruction, an older and wiser Gillies reflected that this procedure probably would not have been successful. But he added that, at the time, "it was the most thrilling thing I had ever seen. I fell in love with the work on the spot."

Gillies was fortunate to fall in with inspiring figures at the postings he was given. A short while later, he left the Val-de-Grâce to take up his new assignment at the Allied Forces Base Hospital in Étaples. It was around this time that Gillies first came into contact with Varaztad Kazanjian, an Armenian American dental surgeon working with the No. 22 General Hospital in Camiers. Kazanjian had traveled to France in June 1915 with a volunteer contingent of medical personnel from Harvard University to offer assistance to the British forces. It wasn't long after arriving that Kazanjian set up a special unit for the treatment of soldiers with maxillofacial injuries. Like Valadier, he spent his days and nights triaging men with broken jaws, smashed noses, and facial lacerations. Gillies admired Kazanjian, whose conscientious work underlined for him the primary importance of dentistry in facial reconstruction.

"[H]is use of weighted dentures produced such soft lips and ample chins," Gillies marveled. The two men would remain in close contact throughout the war.

Gillies was determined to put his newfound knowledge and his enthusiasm for plastic surgery to use. Although Kazanjian, Valadier, and Morestin were working hard to address the war's growing number of facial injuries, Gillies knew that the current system under which these soldiers were being treated was fragmented and inefficient. While some men were lucky enough to fall under the care of these specialists, most were sent to general surgeons who hastily patched up patients before sending them back to the trenches to fight.

Complicating the problem was the fact that plastic surgery was not yet a formal discipline, and virtually no British surgeon had any clinical experience in this field. Gillies believed that if casualties were directed to, and treated in, a specific location, surgeons working there could learn more quickly on the job and thus treat subsequent influxes of patients more effectively. What was needed, he thought, was a hub where a variety of practitioners could come together to address the staggering number of facial injuries being inflicted on the Western Front. In this way, surgical methods could be tried, tested, and standardized.

At the end of 1915, Gillies took it upon himself to approach Sir William Arbuthnot Lane, who was the senior surgeon in charge of the Cambridge Military Hospital. Named after one of the dukes of Cambridge, this facility was actually at Aldershot, in the county of Hampshire. Lane—a forward-thinking man who had been an early advocate of sterile caps, masks, and gloves in the operating room—was exactly the right person to contact. He had recently helped concentrate amputees at Queen Mary's Hospital in Roehampton so that they could receive specialized care. Lane was therefore receptive to Gillies's vision of a unit dedicated entirely to face and jaw wounds at Aldershot, and he knew that Gillies was well placed to take charge, given his experience working on similar cases in France. He took Gillies's idea to his good friend Alfred Keogh, Director General of the British Army Medical Services. It wasn't long before the request was granted.

On January 11, 1916, Gillies received an order from the War Office to report to the Cambridge Military Hospital for special duty in connection with plastic surgery. Now he just needed to prove he was equal to the task.

⇥ 4 ⇤

A STRANGE NEW ART

t was a cold January day in 1916 when Harold Gillies strolled onto the ward at Aldershot's Cambridge Military Hospital for the first time. He looked around proprietorially, and then turned to an Irish nurse named Catherine Black. "This is the very thing for [my jaw cases] . . . these poor fellows [here] can be transferred to other wards." Black, who had just settled into the rhythm of her day's work, was taken aback by the nonchalance with which he delivered this order, as if finding new beds for seriously wounded men in an already overcrowded hospital was the work of a few minutes.

Black would soon learn that the word "impossible" was not in Gillies's vocabulary. It was futile to argue with him. In his view, what needed to be done simply *would* be done. "He would not admit defeat," Black observed. And that attitude would serve his patients well as he began the herculean task of rebuilding their faces.

Black, who was in her late thirties, had been working at the Royal London Hospital when the First World War broke out. When she volunteered for service as a nurse, she didn't actually believe she would ever be called up. Like most people, she thought the conflict would be over by Christmas, before her skills would be required. As time passed and the number of casualties grew, however, so too did the need for experienced doctors and nurses. Before she knew it, Black was donning a military nurse's uniform similar to a nun's habit. It consisted of a simple gray dress, a shoulder cape with a scarlet border, and a white muslin cap. By the time Gillies traipsed onto her ward in 1916, she was beginning to wonder if the war would ever end.

The Cambridge Military Hospital had a long central corridor that extended nearly as far as the eye could see. It had been built in line with Florence Nightingale's sanitation reforms of the nineteenth century, with spacious wards illuminated by tall windows on either side to enable cross-ventilation. The building, which was neoclassical in style and featured a central clock tower, sat atop a hill—a spot carefully chosen by the original architects, who believed this lofty location would allow the wind to "sweep away" infections. But when Black arrived on the scene, the vast hospital was in total disarray.

Almost all the regular army nurses had been shipped off to France a year earlier, taking with them the best medical equipment. Most of the hospital's surgeons had also been sent to serve in facilities close to the front, leaving behind elderly staff unfit for military service or inexperienced doctors fresh out of medical school. In time, even these novices would be sent abroad. The American nurse Ellen La Motte summed up the bleakness of the situation when she observed, "[A]ll those young men who did not know much, and all those old men who had never known much, and had forgotten most of that, were up here . . . learning. This had to be done, because there were not enough good doctors to go round . . . it was necessary to furbish up the immature and the senile." So severe was the shortage of medical staff on the Continent that on at least one occasion, a regimental doctor killed in the line of duty was replaced by a veterinarian.

Back in Aldershot, the "motley crowd" (as Black called them) of medical veterans and neophytes constituted the last hope for men arriving at the Cambridge Military Hospital. Many casualties were already completely debilitated from their life in the trenches. Although each man was supposed to be pinned with a label indicating his name, number, regiment, type of wound, and a note stating whether he had received an anti-tetanus injection, many bore labels that simply read "GOK" (God only knows).

The sight of a soldier with injuries to his head and neck could send shock waves through the ranks of the most battle-hardened nurses. A Swiss nurse, Henriette Rémi, working in a German hospital during the war recalled disfigured men with "mutilated debris in lieu of faces." Mary Borden—a nurse who later suffered a mental breakdown—described a time when she lifted the bandage from a soldier's head, and half his brain slipped out. Such carnage brought home to the nursing staff on the wards the severity of the distant conflict raging on the front. "You could not go through the horrors we went through, see the things we saw and remain the same," Catherine Black later reflected. "You went into it young and light-hearted; you came out older than any span of years could make you."

After Harold Gillies issued the order to empty the ward, Black began the logistical nightmare of transferring the men under her care to different sections of the hospital. Meanwhile, Gillies—who had only just been reunited with his family—bade his wife and two young children farewell once more and returned to the Continent with hopes of gaining further experience in reconstructive surgery. His travels eventually brought him back to Paris, where he decided to pay a second visit to Hippolyte Morestin at the Val-de-Grâce.

Rumors were circulating that the French surgeon had been conducting successful cartilage grafts on patients. "My tongue was literally hanging out from the thirst of knowledge that I hoped to obtain by observing this surgeon's work," Gillies wrote. But when he arrived, Morestin unexpectedly denied the eager would-be student entry to the operating room. In desperation, Gillies shoved his official permit into Morestin's hand, only to watch with dismay as the elder surgeon shrugged his shoulders and walked away. Gillies was one of countless foreign surgeons who turned up at the hospital seeking the Frenchman's advice during the war, which may explain the cold reception he received. Morestin's student Georges Duhamel later recounted that "[w]e ought to have treated [these visitors] as guests . . . This was certainly not the rule."

On his return from France, Gillies headed straight to the War Office in London. This neobaroque edifice of Portland stone was located on Whitehall's Horse Guards Avenue in the heart of the city. Alongside four elaborate domes on its roof were the silent, sculpted personifications of Truth and Justice, Victory and Fame, and—unsurprisingly, given the building's function—War and Peace. The building's one thousand rooms and seven floors of offices and corridors vibrated with nervous activity when Gillies arrived.

During his visit he suggested that soldiers with facial injuries be tagged with labels directing them to his new unit at the Cambridge Military Hospital. But no one there seemed particularly receptive to this idea. After all, as office-bound functionaries, they hadn't witnessed firsthand the wounds Britain's soldiers were suffering, and they may not have considered this a pressing concern. Gillies recalled the dismissive attitude of his superiors: "I was told by the War Office to 'Run away little boy. We are far too busy for that sort of thing.'"

Gillies persisted, knowing that if the job were to be done, he would have to do it himself. He strode onto the Strand to find a stationer's shop and spent £10 of his own money on labels, which he proceeded to address to himself at Aldershot. He then returned to the War Office and asked that the labels be sent to the front for soldiers with facial wounds. With this irregular request lodged, he left the building, hoping that the labels would find their way to the right patients. Gillies was ready to get to work.

On his return to Aldershot, Gillies immediately began to recruit members of his surgical team. Unlike Morestin, he sought to surround himself with a variety of practitioners with whom he could collaborate. Auguste Charles Valadier had taught him the importance of dental techniques in reconstructing jaws, and Gillies had concluded early on that the restoration of the face's substructure was just as important as repairing the soft tissues. "In no other part of the work does cooperation of the dentist and surgeon come more fully into play" than when rebuilding shattered jaws, he reported in *The Lancet* shortly after he began work at Aldershot. "Disappointment is in store for him who would confine his repair to the surface tissues, heedless of Nature's lessons in architecture."

The dentist's primary role in facial reconstruction was to ensure that the patient would be able to eat and speak with relative ease, and in this regard, he was essential to the overall success of reconstructive work. Due to the high risk of infection in the pre-antibiotic era, there was no safe way to stabilize fractured bones internally with metal plates, rods, wires, or screws, as a dental surgeon might do today. Instead, frames and pins had to be applied *externally* to the face in order to secure the jaw while the surgeon addressed other aspects of the injury.

Given these challenges, it's hardly surprising that when the time came, Gillies requested that not one but two dental surgeons join him, in addition to an anesthetist and a surgical assistant. In time, his team would grow to include other surgeons, radiologists, artists, sculptors, and photographers—reflecting Gillies's belief that facial reconstruction required multidisciplinary care.

Meanwhile, in a small but heartening miracle, Gillies's simple trick with the stationery had actually worked. Within weeks of his visit to the War Office, injured men began arriving with his hand-written labels pinned to their tattered uniforms. Before long, the ward was swamped with casualties in desperate need of medical attention. "Whenever the head of a careless soldier . . . peeped out of the trench and a ray of moonlight touched his white face, there was another patient for [us]," he wrote.

Due to pressures at the front, most surgeons simply stitched together the edges of a gaping wound and hoped that nature would do the rest. Inevitably, this caused problems for Gillies once these men were placed in his care. He understood the instinct to take this approach when having to repair faces with large chunks of flesh missing. "It seemed that the first important step was to cover the gap," he wrote, "and therefore it was a temptation to close the hole by pulling adjacent tissues together." Unfortunately, this hasty method often led to cellular destruction, or necrosis, of the affected area, which could spread to nearby healthy tissue if left untreated.

There was no easy solution when it came to addressing tissue loss in the face. Before the war, surgeons had periodically experimented with implanted devices, such as metal and celluloid plates, despite the risk of infection. Some doctors even injected hot paraffin wax into the face in order to contour, repair, or perfect a patient's features. In the early 1900s, Gladys Deacon, Duchess of Marlborough, allowed a doctor to inject wax into the bridge of her nose in hopes of achieving a perfect Grecian profile. As she aged, the wax shifted, settling in her chin and causing lumps to form. She was so upset by her changed appearance that she eventually became a recluse, locking herself away from the world when she could no longer conform to the beauty standards of her day. Given the myriad complications that could arise from implanting foreign substances in the face—not least among them the likelihood of rejection—Gillies preferred bone for bone, cartilage for cartilage, skin for skin.

The consequences of hasty, ill-conceived repairs on the battle-field were never more evident than in the case of one man whose lower lip and middle section of jaw had been sheared off by a bullet. Due to the severity of his wound, he was rushed to a field hospital, where surgeons frantically knitted together the pieces of his broken mandible, resulting in a badly scarred lip that was drawn downward in a "viciously disfiguring way." Eventually, the man was sent to Aldershot. There, Gillies had to reopen the original wound in order to enlarge it. Afterward, he borrowed tissue from an area below the man's jaw to create a skin flap, which he then rotated to cover the defect.

The term "flap" originated in the sixteenth century and comes from the Dutch word *flappe*, meaning something that hangs broad and loose and is attached on only one side. In surgery, a flap is a section of healthy tissue that is partially detached from its original site and moved to cover a wound. The flap has its own blood supply in the form of a single large artery or multiple smaller blood vessels. There are two types of flap: local and distant. In the former, skin can be transferred from a site adjacent to the wound by cutting a flap from an intact piece of tissue and rotating it about its pedicle—the point where the flap remains attached to the body—to cover the injury. The incisions made by the surgeon are then stitched closed. Alternatively, a flap can be advanced in a direct line; that is, intact skin is pulled taut to cover a wound before being stitched into place.

By contrast, distant flaps involve the transfer of tissue from a remote area on the body. But, as with the local method, one side of the flap has to remain connected to the original site in order to maintain its blood supply. Performing this procedure could entail leaving a long "tether" of skin—its underside comprised of raw flesh—exposed. This left the flap susceptible to infection, which could prove life-threatening. In time, Gillies would find a solution to this problem by inventing a new kind of flap that would dramatically reduce the risk of infection. In those early days at Aldershot, however, he had to manage the situation as best he could.

Many cases presented unforeseen challenges. One young man's jaw was crushed by being hit with shrapnel. As a result, blood continuously trickled into his airway until it overflowed from his nose and the corners of his mouth. Remarkably, he avoided suffocating by sitting upright on a ten-day journey back to Britain. Because of the nature of his wound, however, he couldn't eat or drink effectively. Consequently, he arrived at the Cambridge Military Hospital severely malnourished and dehydrated, with his face "bathed in pus, foul-smelling and . . . gangrenous."

Gillies's first task was to irrigate the infected wound, as he had seen Valadier do countless times at the specialist unit in Boulogne. Afterward, the nursing staff periodically flushed out the wound while the patient sat upright with a kidney basin under his chin to catch the fluid. Over the course of several days and weeks, Gillies and his team were able to halt the infection and promote healing in the damaged tissues. Only then was he able to turn his attention to the reconstructive work necessary.

Infections weren't the only problem Gillies faced at Aldershot. Countless other challenges presented themselves at every turn. Edward D. Toland, who worked at a field hospital based in the Hotel Majestic in Paris in 1914, described in excruciating detail the difficulties the staff faced with a patient who couldn't eat solid food due to the nature of his injuries:

> [The patient] has to lie face downward and of course cannot take anything but liquid food. [We] put a basin in front of him and a rubber cloth around his neck; then he pushes a rubber tube down his throat and we pour in beef tea, or

milk, through a funnel. About every other swallow, it goes down the wrong way and he strangles for two minutes; then nods his head as if to say all ready again.

Gillies understood that it was crucial to maintain the men's strength so that they could endure a series of grueling operations. Therefore, at Aldershot, specific attention was given to their diet, which mainly consisted of soups, milk, and artificially digested foods. Liquid meals were fed to patients through rubber-tubed feeding cups, and their mouths and throats were then doused with water to keep them clear of infection. For those who could chew, there were also eggs—and lots of them. According to Nurse Black, she and her colleagues might serve as many as three hundred eggs to patients in a single day. The work was endless. No sooner had the staff finished feeding the men than they had to begin again.

Thus, the first faltering steps of establishing a specialized jaw unit gradually settled into a routine at Aldershot.

As winter dragged on, military medical personnel worked overtime to return wounded soldiers to the battlefield in order to deal with the shortage of men being generated by the world's first large-scale industrialized war. A reporter at *The Times* remarked that the injured "began to have a potential value when new recruits were difficult to obtain in adequate numbers."

The demoralizing pressure to send men back to the front was not one that Gillies could altogether escape. He bemoaned the fact that "as soon as the healing had occurred the soldier was sent back to his battalion or battery, often looking like a travesty of his former self." As soon as his boys could shoulder the weight of their field kit, they were sent back into battle, only to be torn to pieces again. One war nurse wondered, "Was it not all a dead-end occupation, nursing back to health men to be patched up and returned to the trenches?"

In an address to the Medical Society of London, Gillies acknowledged this dilemma. "I would have you know that my first duty is to the Army, and that this involves the sending back to duty as many soldiers as possible in the shortest time," he wrote. "[M]y second obligation is to the patient and to do the best for him that in me lies, whether he is to be a spectacular success or merely a poor, patched-up pensioner; and my third duty is to contribute as freely as possible to science and to the knowledge of surgery." Of course, these three duties frequently clashed, and Gillies admitted that pressures from the front often greatly hampered his duties to his patients and his profession.

Ironically, medical advances that owed their existence to remediating the horrors of the war also served to prolong it. One soldier remarked, "A casualty was not a matter for horror but replacement." As doctors rushed to patch men up, they were inadvertently feeding the war machine with more manpower once their patients were rehabilitated. The war surgeon Fred H. Albee observed, "There could only be one bright spot in this deplorable result—that in the long run, humanity would benefit from the knowledge surgeons had gained in time of war." But while the war was raging, there was only one focus: return as many soldiers to the front as possible and as quickly as practicable.

While Gillies labored hard to mend his patients at Aldershot, the British Army began to develop measures to protect infantrymen from head injuries. During the first year of the war, soldiers wore fabric caps and went into combat without purpose-designed protective headgear of any kind. These caps failed to safeguard troops not only against bullets and shrapnel, but also any volatile weather conditions on the front. "Nothing could have been devised more unsuitable for active service than the present military cap," a writer for *The Times* bemoaned. "[It] makes the wearer easily discernible . . . [and provides] no protection against sun or rain." Their French and Belgian allies shared the same risks. Even the Germans' leather *Pickelhaube* with its characteristic brass spike offered little resistance against a shell splinter traveling at high velocity.

In 1915, the French army constructed the first metal headgear. It was a bowl-shaped piece made from soft steel that sat underneath a soldier's cloth hat. Unfortunately, if shell splinters pierced the skullcap, they carried pieces of dirty cloth into the wound, increasing the risk of infection. To make matters worse, these flimsy metal caps shattered into slivers of shrapnel when hit, inflicting further damage on the wearer and those near him. The French went back to the drawing board and came up with the Adrian helmet, a dome-shaped headpiece with a narrow brim that could be worn in lieu of a fabric cap—a design credited to Intendant-General August-Louis Adrian. Despite these improvements, the British War Office remained unimpressed and sought a better solution that could be swiftly manufactured and distributed to their own troops.

Later that year, the British Army selected a design patented by John Brodie. The Brodie helmet (as it became known) was cut from a single sheet of Hadfield steel, a material known for its high impact resistance. It was then pressed to form a "soup bowl" shape, which provided a two-inch-wide brim for additional protection. Not only was the Brodie helmet strong, but it was also simpler and quicker to produce than the French design. Given the growing number of head wounds on the Western Front, high-speed production was essential. By the summer of 1916, one million Brodie helmets had already been distributed to troops at the front, and it soon become another iconic symbol of the war. It was the first helmet to be given to *all* soldiers serving in the British and Commonwealth armies, regardless of rank.

The success of the Brodie helmet prompted some to wonder whether steel protective gear should extend beyond the head to other parts of the body. In an article for *The Times*, one reporter wrote: "we should require breast and body pieces with knee and elbow caps" for men fighting in the trenches. This idea was never given serious consideration, however, probably due to the cost of implementing it, as well as the impact such heavy armor might have on mobility within the trenches and on the battlefield.

Attempts to improve the helmet also included the addition of a chain-mail veil, which was tested against a three-ounce shrapnel round fired from one hundred yards. Men inside tanks found that the veil provided effective protection against "splash," or flying metal splinters caused by the impact of bullets hitting the outer steel of the tank's body. But most infantrymen felt that the veil was too distracting and removed it from their helmets, thus defeating its purpose. Other more practical inventions followed, including periscope sights on rifles that allowed the shooter to remain concealed.

And yet, despite these attempts to prevent head injuries, the face remained largely exposed and vulnerable, and the wounded continued to stream into Harold Gillies's ward by the hundreds during those first months at Aldershot.

From the outset, Gillies demonstrated an extraordinary ability to see past a soldier's disfigurement. Those who knew him saw "a man of steel nerve and a great heart" who viewed his patients as more than just numbered combatants. D. M. Caldecott Smith, whose brother was under Gillies's care at Aldershot, remembered the doctor as being "full of human kindness." Similarly, Sergeant Reginald Evans expressed astonishment that "ordinary soldiers received as much care as officers." He wrote that Gillies "even dressed my wounds himself and visited me at night to see if I was comfortable, though he was up to his eyes in work." Evans attributed the relative normalcy of his later life to Gillies's successful reconstructive work: "I owe much of my happiness to him."

Unsurprisingly, the emotional impact of a facial injury on a soldier could be extreme. The surgeon Fred Albee noted that the "psychological effect on a man who must go through life, an object of horror to himself as well as to others, is beyond description." He observed that a disfigured soldier often felt like a "stranger to his world," adding that it must be "unmitigated hell to feel like a stranger to yourself."

For his patients, Gillies's very presence had its own curative power. He would often comfort the wounded with his trademark reassurance: "Don't worry, sonny . . . you'll be all right and have as good a face as most of us before we're finished with you." Gillies's easy manner and sense of humor rarely failed to lift moods on the wards. "I thanked Heaven for an inherited ability to twist fun out of the ordinary things of life," he remarked. Throughout his time at Aldershot, his patients came to love him for it.

Not everyone was as comfortable around disfigured men as Gillies. Even for experienced medical personnel, encountering someone with a facial injury could be traumatic, and their reactions could cause the patient even more distress. Ward Muir, who worked at the 3rd London General Hospital in Wandsworth, was surprised by his reaction. "I never felt any embarrassment in . . . confronting a patient, however deplorable his state, however humiliating his dependence on my services, until I came in contact with certain wounds of the face," he later confessed. Muir imagined each soldier as he must have been before the war—"a wholesome and pleasing specimen of English youth"—which only made matters worse when he was standing before these "broken gargoyles." He feared that he might inadvertently "let the poor victim perceive what I perceived: namely, that he was hideous."

Gillies, however, inspired confidence in both his patients and his staff, even in the most desperate situations. "He would set to work on some man who had had half his face literally blown to pieces with the skin that was left hanging in shreds, and the jaw-bones crushed to pulp that felt like sand under your fingers," Nurse Black remembered. Her admiration for Gillies grew all the while she was stationed there. Although the work was harrowing, Black quickly came to recognize the groundbreaking importance of what he and his team were doing. "[T]he Great War in which millions of lives were sacrificed was indirectly responsible for saving millions of others," she shrewdly observed.

But progress was slow in the early days, and Gillies did not always share Catherine Black's confidence in his own skills. "This was a strange new art, and unlike the student today, who is weaned on small scar excisions and gradually graduated to a single harelip, we were suddenly asked to produce half a face," he later recalled. With no textbooks to guide him and no teachers to consult, Gillies had to rely on his imagination to help him visualize solutions to the problems set before him. And yet, the sheer scale of the catastrophe provided a unique opportunity for plastic surgery to evolve and for best practices to become standardized. Years later, Gillies would look back on his days at Aldershot with wonder. "All this time," he wrote, "we were fumbling towards new methods and new results without the boon of sulfa drugs, plasma or penicillin."

Gillies experienced as many defeats as he did victories, with each new blow feeling just as acute as the one preceding it. The death of one Private William Henry Young was difficult for him to bear, especially given the heartbreaking circumstances under which the soldier had arrived at Aldershot.

Young was a portly man with a handlebar mustache and chubby cheeks that made him look younger than his forty years. Early in the war, he was shot in the thigh by a sniper. After a lengthy recovery, he returned to duty, only to fall victim to a chlorine gas attack in the spring of 1915, not far from where Gillies was working at the Belgian Field Hospital. The incident left him with impaired vision. That winter, Young was sent back to the front, just east of Foncquevillers in northern France, where another catastrophe was about to befall him.

As dawn crept over the trenches on a cold and wet December morning, just days before Christmas, Young spotted a wounded officer lying 150 yards away. Acting without orders amid heavy enemy fire, he made his way through the barbed wire and crawled to where Sergeant Walter Allan lay bleeding. The officer ordered Young back to safety, but he would not be deterred. Young gathered the injured man up and began making his way back to the trenches. Just as he was nearing safety, a bullet smashed through Young's jaw, and another hit him square in the chest. Another soldier rushed to help him, and together they pulled Allan back into the trenches, saving the officer's life.

Then, despite the severity of his own injuries, Young walked half a mile to a regimental aid post to receive medical treatment. Eventually, he was evacuated back to Britain, where he underwent several operations on his shattered jaw at a hospital in Exeter. During that time, his platoon commander wrote to Young's wife, Mary: "I am sure that it must be a great consolation to you to know that he received his wounds whilst rescuing a wounded comrade, and we all hope that he gets from the officials the recognition he deserves. From us he has our lasting admiration." It was not long after the letter was sent that the British Army awarded Young the Victoria Cross, the military's highest honor. Young wrote to his wife from his hospital bed, "I am naturally very proud of the great honour, both for my sake, and for the sake of you and the kiddies."

A fund was established to support Young and his family during his recovery. The public gave generously, and more than £500 was raised. Young was granted leave to return home to Preston, where he was welcomed as a hero. Waiting for him at the train station was a horse and carriage sent by the mayor to convey him to the town hall, where further celebrations awaited. In a spontaneous display of admiration, local volunteers of the East Lancashire Regiment unharnessed the horses and pulled the carriage themselves through the cheering crowd, while a band played "See the Conquering Hero Comes."

As the celebrations wound down, Young turned his attention back to his recovery. Though the wound in his chest was not serious, his facial injury meant that he was unable to consume food except through a tube. It was decided that he should be sent to the new hospital unit at Aldershot.

Gillies was confident in his ability to repair Young's jaw, but the operation was not without risk. One of the greatest challenges his team faced was the administration of anesthesia. Facial injuries often resulted in the obstruction of the airway, either from swelling of the tongue and throat or from the loss of muscles controlling the larynx. As stretcher-bearers had learned, this situation was especially dangerous when a patient was lying on his back, such as he typically was under anesthesia.

By the time Young arrived at Aldershot, Gillies was all too familiar with this problem. He once had to help an anesthetist pull a patient's tongue forward in order to clear the airway after the man turned blue from lack of oxygen. "I washed up and continued to chisel the donor bone graft, when the patient went bad again, and the anesthetist once more had to be helped to keep the patient alive," Gillies wrote.

While placing a patient in a seated position was one solution, this, too, presented challenges, since it could cause a drop in blood pressure. Worse still was the risk that the patient might wake up in the middle of the procedure. Gillies remembered a man named Morrison who had done just that. "It was not a little disconcerting when near the end of the operation he began speaking to me in French and German," Gillies recalled. "I was trying to answer back as well as help hold him up, fix his graft into position, [and] suture his wound." A day never passed without some new and unforeseen challenge.

Private Young was anxious about his upcoming operation when

he arrived at the Cambridge Military Hospital. He told Gillies that the chloroform used on him in the past had left him in a bad state. Indeed, military surgeons discovered in the early years of war that chloroform, a common anesthetic at the time, could cause ventricular fibrillation, a fatal cardiac arrhythmia. This effect was especially dangerous in men like Young whose health was already compromised. Still, Gillies pushed ahead with the surgery, hoping—as he always did—for the best outcome. On the day of the procedure, he used as little anesthesia as possible. Nonetheless, Young slipped into a coma. Despite Gillies's repeated attempts to revive him, he never regained consciousness. Having survived multiple gunshot wounds and poison gas, Private William Henry Young was dead from sudden heart failure. He left behind nine young children and a wife.

Gillies was devastated. He wrote to Young's widow, remarking that he was "utterly miserable" due to the fatal outcome of the operation, lamenting that "[i]t seems so terribly cruel to go through all he did, and so well, and then to die through the worst of bad luck." He assured her that every precaution had been taken. "Everything that could be done was done for him," Gillies wrote, "and five doctors saw him and did what was possible." Nonetheless, the failure weighed heavily on his mind. Solving the problem of how to safely administer anesthesia to patients with facial injuries would become a priority for Gillies and his team.

Young's body made the two-hundred-mile journey back to Preston for burial. Crowds of mourners gathered outside the hero's home to see his coffin carried through the streets to the local cemetery. A contingent of fifty wounded men, including two from Aldershot, made the journey north to attend the funeral.

But however deeply Gillies mourned Young's passing, the overwhelming number of soldiers requiring his help prevented him from paying his respects. It would not be long before the deluge of patients at Aldershot became unmanageable, even for a surgeon as dedicated and focused as Harold Gillies.

⊰ 5 ⊱

THE CHAMBER OF HORRORS

I t was the spring of 1916, and the first buds of the season had
begun to bloom. Gillies watched from a window as a "crew
of wobbly jaws and half faces" filed out of the hospital for a
walk. He knew that his patients were excited about the opportunity to escape the confines of the building for a short while and
take some fresh air. But he worried about the bank of heavy gray
clouds gathering in the near distance. At the slightest hint of rain,
he knew that the sergeant-in-charge would call a halt to proceedings, forcing "the sad column of bandages [back] into the dismal
wards."

As Gillies turned to resume his rounds, he could hear the men
outside singing a familiar tune: "It's a long way to Tipperary, it's a
long way to go!" Under different circumstances, a cheerful gaggle
of soldiers might have been a welcome sight to members of the
surrounding community. But in their current injured state, they
only served as a stark reminder of the violence being waged on the
front. Anxious mothers called their children in from play at the
sight of the war's human wreckage, while others offered up silent

prayers for their own brothers, sons, and fathers who were so far from home.

In the distance, the dark clouds grew ever more ominous.

It was during this time that Gillies reflected on the lessons he had learned since beginning his work at Aldershot. Months of trial and error had taught him that reconstructive surgery had to be carried out incrementally. He bemoaned the fact that a "majority of plastic operations are unavoidably long; the insertion of sutures alone is apt to occupy a skilled surgeon more than half an hour." Not only that, but sometimes a single patient needed as many as fifteen operations, and these had to be spread out over an extended period. "Surgical haste definitely led to irrevocable waste of tissue," Gillies observed. His mantra quickly became: "never do today what can be put off till tomorrow."

It wasn't long before Gillies's work began to attract the attention of the national press. A reporter from the *Daily Mail* visited Aldershot and wrote of the sobering experience. "Nowhere does the sheer horror and savagery of modern warfare appeal so vividly to the mind and senses as in a tour of these wards," he told his readers. The journalist, who described the hospital as "[s]omewhere in England," keenly observed that "[a] shattered arm excites our pity, an absent leg arouses our compassion, but a face ravaged by shrapnel . . . cannot fail to arouse a certain amount of repulsion." One cannot help but wonder what the patients themselves may have thought when they encountered such dehumanizing descriptions of their injuries in the news.

Still, on a happier note, the same journalist went on to praise Gillies and his team for their tireless efforts to repair the faces of disfigured soldiers, noting that "there is to be found in these wards a supreme instance of triumph of the human spirit." It was the first published account of the work being done at Aldershot, and Gillies later acknowledged the benefits of publicity in those early days. "It did a great deal of good," he reflected. "Only thus, by explaining to the public and to the general practitioner what can be done by a new branch of surgery, are sufferers able to benefit from it."

As Gillies's surgical unit gained a foothold, he recognized the importance of documenting his pioneering work so that surgeons could replicate it in the future. Frustrated by the difficulty of explaining his innovative techniques in words, he decided to enroll in correspondence courses at the Press Art School in London so that he could learn how to draw his patients. His instructor said that "Gillies had an air of quiet efficiency and of being very sure about what he wanted to do." As with sports, Gillies soon discovered that he had a natural aptitude for drawing. But however accomplished Gillies's sketches were, greater challenges awaited him at Aldershot. Each case would demand his undivided focus to address the unique surgical challenges it posed. He would have little time for sketching.

A well-timed recommendation from his friend Bernard Darwin, a journalist at *The Times*, led Gillies to the "great Henry Tonks," who held an administrative post at the Cambridge Military Hospital. Tonks also happened to be a teacher at the Slade School of Fine Art in London, and his was exactly the kind of talent that Gillies was looking to add to his ever-growing team.

Henry Tonks was a formidable presence. At six foot four inches, he loomed over his pupils like a question mark, addressing them in "cold discouraging tones." Helen Lessore described her professor as "lean and ascetic looking, with large ears, hooded eyes, a nose dropping vertically from the bridge like an eagle's beak and quivering camel-like mouth." He was ruthless in his criticism and told his students that tolerating bad drawing was like living with a lie. The English painter Gilbert Spencer felt like flinging himself under a train after receiving a critique of his artwork from Tonks. Spencer's brother Stanley, also an artist, told him to pay Tonks no mind. The man was critical of every student who passed through his classroom, without exception.

But underneath his seemingly tough exterior was a sensitive soul attuned to the suffering of others. "I used to feel the illness of anyone I loved so acutely as to make it almost unbearable," he once wrote. His sensitivity toward those in pain would be tested to its limit when the war broke out. Another side to his character was also seemingly at odds with his somber manner, though not with his talent for capturing a likeness. He nurtured a private passion for humorous illustration and frequently drew accomplished caricatures of his friends.

Given that he was fifty-two, Henry Tonks was past fighting age when the first guns boomed over the fields of Europe. Unlike many of his contemporaries, he knew early on that the conflict would not end quickly. The revelation struck him like a thunderbolt one warm summer evening in August 1914, as he made his way through London for supper at the home of the novelist George Moore. On his journey, he caught sight of posters plastered all over the city announcing the fall of the Belgian stronghold Namur. "I bring grave news, my friend," he said as Moore ushered him into the spacious drawing room of a charming Georgian house just off Ebury Street, in the swanky district of Belgravia. Not wishing to have the mood of the party killed before it had begun, Moore airily replied, "Oh, it all makes the newspapers more interesting." Tonks was unamused by his friend's dismissive attitude and made sure Moore knew it. Later, the novelist would write of the artist: "his seriousness is depressing, but I am inclined to think that I would rather be depressed by Tonks than amused by any other man."

It wasn't long after this incident that Tonks decided to volunteer at a camp in Dorchester, on the southwest coast of England. Ultimately a POW camp, it had been established just ten days after the war began, when the British government started rounding up thousands of expatriates from enemy nations to be detained until the war's end. Chief among them were German immigrants.

After the Russians, Germans were the second-largest migrant group living in London, according to the 1911 census. The expatriate community was serviced by a dozen German churches, a German hospital, and two German-language newspapers. German merchants, barbers, bakers, and other tradespeople were essential to the city's economic life. Germans made up 10 percent of London's restaurant waitstaff, and a German governess had become a feature in a great number of the capital's wealthy households.

When the war broke out, however, many of these same workers found themselves driven from their jobs. No German, no matter how important, was immune to the hostilities occasioned by war. Prince Louis of Battenberg, First Sea Lord of the British Admiralty, was forced out of his post after the press led a campaign against him, calling for his dismissal. The English journalist Horatio Bottomley declared that it was "a crime against our Empire to trust our secrets of National Defence to any alien-born official." Not long after Louis's resignation, he also relinquished his royal title and anglicized his family name, changing it from "Battenberg" to "Mountbatten." Countless others followed suit, including the British royal family, who abandoned the Teutonic surname of "Saxe-Coburg" for the more acceptable "Windsor" in 1917.

Unfortunately, not everyone could adapt as easily. Some Germans couldn't cope with the stress of suddenly finding themselves "enemy aliens" in the country they called home. When Joseph Pottsmeyer lost his job, he tried to secure another position, but to no avail. Depressed and despondent, he penned a note expressing his admiration for England before hanging himself. Another man, named John Pfeiffer, shot himself in the eye. When that failed to kill him, he lifted the gun to his temple and shot himself a second time.

Anti-German sentiments infected all areas of life. The popular

"German sausage" was renamed "luncheon sausage" by one of Britain's leading grocers. More insidious than the rebranding of food and changing of surnames was the government's systematic deportation of German women, children, and the elderly. Men in their prime were held in internment camps over concerns that repatriating them would prompt them to enlist in the German army. Such camps sprang up all over Britain. The largest, Knockaloe, was established on the Isle of Man and held over 23,000 men at its peak. But one of the first was in Dorchester.

Ten days after Britain entered the war, the camp's first inmates were escorted to their quarters inside the newly opened prison. The group consisted of eight German civilians. As they filed into the building, crowds of curious onlookers gathered outside. Shortly afterward, military prisoners also began arriving from the Western Front. Many of these men were wounded and required medical attention. Dr. W. B. Cosens oversaw the camp's hospital, which was equipped with an operating room. It was to Cosens that Tonks wrote to offer his services.

Tonks's decision to volunteer at the prison camp wasn't as illogical as it might seem. He was not only an acclaimed artist at the onset of war. He was also a qualified doctor. In 1886, he took a post as a house surgeon at the London Hospital, the same year that Joseph Merrick moved onto a ward as a permanent resident. Merrick, who was severely deformed due to a mysterious condition (possibly neurofibromatosis), had been dubbed the "Elephant Man" because of his enlarged cranium and the spongy skin hanging from his face. He had been performing as a curiosity in sideshows when the surgeon Frederick Treves discovered him and invited him to become a patient at the hospital. Through Merrick's demeaning experiences, Tonks would have been well aware of the stigma that marred the lives of the disfigured.

During his medical residency, Tonks enrolled in evening classes at the Westminster School of Art. After a short while, he gave up surgery to pursue art professionally—much to his father's disappointment. George Moore said of his friend's decision: "the desire of art must have been strong in Tonks, for it obliged him to abandon the career he had chosen and in which he was successful, for another in which he might have been a failure."

Luckily for Tonks, failure was not in his future. On his departure from the London Hospital, a medical student stopped him in the corridor and asked to purchase two of his watercolor paintings. "I had no banking account at the time and walked a very long distance into the East End of London to cash the cheque at his bank, and came back with twenty-five golden sovereigns in my pocket"—as well as a new confidence in his prospects.

A telegram from Tonks preceded his arrival at the POW camp in Dorchester. It read, with characteristic pithiness, "coming tomorrow." Cosens tasked him with compiling a list of the various types of wounds that presented on the wards. Instead of producing written descriptions, however, Tonks spent hours meticulously drawing each injury for posterity. He stayed for several weeks before departing as suddenly as he had arrived after giving customarily curt notice. "I am leaving to-morrow." From there, he went to Hill Hall, an officers' hospital in Essex. Cosens didn't hear from the artist again for months.

As more bad news flooded in from the front, Tonks decided to set artistic pursuits aside and serve with the Red Cross. By January 1915, he was stationed at an evacuation hospital in Haute-Marne, France. There, he met the sculptor Kathleen Scott, widow of the ill-fated Antarctic explorer Robert Falcon Scott. She organized and led the hospital's small ambulance service and in time would come to work with Harold Gillies, creating plaster casts of his patients' broken faces to aid in the surgical process.

Tonks was appalled by the carnage he witnessed at the evacuation hospital. "The wounds are horrible," he wrote, "and I for one will be against wars in the future, you have no right to ask men to endure such suffering." It wasn't long before he realized that his medical skills were unequal to the task of dealing with such slaughter. "I am not any use as a doctor," he confessed.

Still, Tonks's sense of duty was strong, and he wanted to serve his country. "A wisdom far above anything we can grasp has taken me in hand," he declared, "and I have got to do what I am told." In 1916, he joined the Royal Army Medical Corps and was sent to the Cambridge Military Hospital in Aldershot—not as a doctor, but as an adjunct secretary. In a letter to a friend, Tonks wrote, "I may have something to do or nothing."

As it turned out, there *was* work for Tonks. And so, on the recommendation of his friend Bernard Darwin, Harold Gillies introduced himself to the renowned artist one afternoon in the spring of 1916. To Gillies's eyes, Tonks looked like "the Duke of Wellington reduced to subaltern's rank," confined as he was to an orderly room and performing mundane administrative tasks. Gillies suggested that Tonks might escape the dullness of secretarial work by joining his surgical team, creating pictorial records of patients before, during, and after their operations. It was the first of many steps Gillies would take to preserve a record of his work. In time, such careful documentation would serve to standardize surgical methods, which would help to establish plastic surgery as a legitimate branch of medicine.

Tonks's anatomical training made him a perfect candidate for the job, and Gillies felt that an artist would be less of an intrusive presence in the operating room than a photographer. Moreover, in an era before color photography had become widespread, Tonks could depict battlefield wounds using a palette of angry crimsons, lurid purples, and moldy greens, capturing the nuances of injured and infected flesh. In this way, his portraits were often more realistic than black-and-white photographs.

As soon as Tonks's position as resident artist was approved by the director of the Cambridge Military Hospital, he began drawing in earnest. He would spend hours in the operating room with Gillies, taking notes and sketching diagrams of complex procedures. "I am doing a number of pastel heads of wounded soldiers who have had their faces knocked about," he wrote to his friend D. S. MacColl, art critic and former keeper of the Tate Gallery in London. "A very good surgeon called Gillies . . . is undertaking what is known as the plastic surgery necessary." Tonks brought an artist's perception to the injuries being treated at the Cambridge Military Hospital. In the same letter, he described a patient who had an enormous hole in his cheek, through which Tonks could see the tongue working. Referring to Diego Velázquez's seventeenth-century portraits of the famously unprepossessing king of Spain, Tonks observed that "[he] rather reminds me of Philip IV as the obstruction to the lymphatics has made his face very blobby."

The nature of his work led Tonks to nurture a kind of intimacy with his subjects. "The medical profession stands alone in giving an observer occasion for a profound study of human beings," Tonks remarked. "Everyone . . . would be the wiser for watching at the bedside of the sick, because the sick man returns to what he was without the trappings he has picked up on his way." Because these portraits were never meant for public display (though they would eventually be exhibited after the war), Tonks felt a new kind of artistic freedom. "When I exhibit a picture [to an audience] I always feel it has lost its virginity," he observed. "In my studio it seems to have a kind of innocence, rather touching, even sometimes a kind of beauty [that diminishes when shown to others]." At Aldershot, he could draw without worrying about the reaction of critics, galleries, and potential patrons. Equally, it seems unlikely that the patients themselves would have given much thought to any public scrutiny the portraits might undergo after their treatment or the war was at an end.

Tonks interpreted much of what he saw in a way that those without his artistic sensibilities might have found perplexing. He likened a soldier whose nose had been severed by a sniper's bullet to a "living damaged Greek head." Yet while Tonks found occasional beauty in his broken subjects, Gillies's unit was, more often than not, an assault on the senses. Tonks described it as a "Chamber of Horrors."

In fact, some of the worst of the injuries to be suffered by combatants were yet to come—as the North Sea became the arena for intensified confrontation between the British and German navies. The omnipresent danger of fire aboard ships would be amplified by the sheer volume of maritime artillery deployed. And the task of picking up the pieces would fall to Harold Gillies, whose work on the victims would revolutionize the nascent field of plastic surgery.

>--<

Sixty miles off the west coast of Denmark, the cold, dark waves of the North Sea were pounding the enormous hull of the British dreadnought *Vanguard*. Belowdecks, Walter Greenaway was waiting for 360 pounds of pillowy dough to rise so he could bake it into bread in the great battleship's ovens. Suddenly, he heard muffled explosions rumbling in the distance. The ship's chief baker quickly wiped his floury hands on his apron before scrambling up to the quarterdeck to investigate the source of the noise. There, Greenaway stood appalled at the ghastly scene before him: "The whole visible horizon . . . was one long blaze of flame." Seconds later, a nearby cruiser was enveloped by fire, jolting Greenaway out of his trance.

It was May 31, 1916, and the Battle of Jutland had just begun. It would not just be the greatest naval battle of the First World War; it would be the largest battleship action of all time, involving 279 ships and over 100,000 men. And it was about to rock the Royal Navy to its core.

Tensions between the British and German navies had been growing steadily since Germany began building up its battle fleet in 1898, sparking a shipbuilding rivalry between the two countries. The competition came to focus on a new class of battleship developed in Britain—the dreadnought. Named after HMS *Dreadnought*, which was completed in 1906, these heavily armored vessels had two revolutionary features: they were equipped entirely with large-caliber guns, and they were powered by steam turbine engines. They rendered all earlier models obsolete overnight. Dreadnoughts soon became a symbol of national power, escalating the arms race between Britain and Germany.

Shortly after the outbreak of war, the British established a naval blockade of Germany, declared the North Sea a war zone, and issued a comprehensive list of contraband that all but prohibited American trade with the Central Powers. By 1916, the blockade was causing serious shortages of food and raw materials, leading to malnutrition and even starvation in the civilian population. Germany was eager to break the blockade, and the British welcomed a showdown, since they believed that their superior numbers and firepower would give them the upper hand in open water. But over the course of thirty-six bloodcurdling hours, that arrogant assumption would be tested to the breaking point.

Just before four o'clock in the afternoon, Vice Admiral Sir David Beatty—in command of the First Battlecruiser Squadron—engaged the Germans' battle cruisers, led by Vice Admiral Franz von Hipper. Thirteen minutes into combat, HMS *Indefatigable* sustained a series of catastrophic shell strikes that pierced its hull. For a British warship long considered almost impregnable, it was a shocking blow. Signaler C. Falmer, who was one of only three men to survive the attack, watched "the guns go up in the air just like matchsticks" seconds before he was thrown clear of the ship, which exploded and sank with 1,017 men trapped on board. The destruction of a warship so early in the battle was a demoralizing blow for the British navy. "There was a colossal double explosion in our line . . . It was an awful sight, even at the distance we were off [on HMS *Lapwing*]; I could clearly see huge funnels, turrets etc flying through the air, while the column of flame and smoke must have been at least 1,500 feet high," recalled Sub-Lieutenant Edward Cordeaux.

Worse was yet to come.

Twenty-five minutes later, German shells from either SMS *Derfflinger* or SMS *Seydlitz* struck HMS *Queen Mary*—the pride of the British fleet—setting off another cataclysmic explosion in this corner of the North Sea. "The [ship] was obliterated by an 800-feet-high mushroom of fiery smoke," Lieutenant Stephen King-Hall recounted long after the terrible event. "As I watched the fiery gravestone, it seemed to waver slightly at the base and I caught a momentary but clear glimpse of the hull . . . sticking out of the water." The doomed *Queen Mary* sank within minutes, killing 1,266 men on board, some of whom were trapped in airtight chambers behind locked doors and hatches. Arthur Gaskin, aboard HMS *Malaya*, felt the tide was turning against his fleet that afternoon: "I realised then that there was death in the air."

Watching from afar, Beatty turned to his men and coolly remarked, "There seems to be something wrong with our bloody ships to-day." When the German High Seas Fleet arrived to join von Hipper, Beatty decided to withdraw his squadron and ordered his ships to turn about. The Germans pursued him, and he led them straight into the path of the entire British Grand Fleet. Beatty's battle cruisers now joined with the rest of the fleet, and the battle continued to rage into the night.

Recognizing that his forces were outnumbered, Admiral Reinhard Scheer ordered the German High Seas Fleet to turn around. Under the cover of darkness, they escaped, depriving the British of a final confrontation the next morning. By that point, the Germans had suffered grave losses: one battleship, one battle cruiser, four light cruisers, five destroyers, and over three thousand men. But the toll taken on the British side was far worse: three battle cruisers, three light cruisers, eight destroyers, and over six thousand men.

The psychological impact was just as great. This Battle of Jutland was imprinted on the mind of every sailor who witnessed it, whether of high rank or low. George Wainford of HMS *Onslaught* noticed a haunting sight: hundreds of dead fish floating on the choppy surface of the North Sea. "I think they'd been killed by concussion," he later recalled. Even the future King George VI, who was on board HMS *Collingwood* when the battle began, was not immune to the intensity of the conflict. "I . . . feel very different now that I have seen a German ship filled with Germans and seen it fired at with our guns," he wrote in a letter back home, reflecting that although it was a "great experience," it was also "one not easily forgotten."

In time, the battle would be seen as an important victory in Britain. The German fleet did not engage with the Royal Navy again and rarely left port for the rest of the war, turning instead to the tactics of submarine warfare. And yet, the British navy had taken a severe beating. An American journalist summed it up by writing that the German fleet had assaulted its jailer but was still behind bars.

Back aboard the Vanguard*, Walter Greenaway discovered that his* bread had continued to bake during the heat of the battle and turned out "fairly creditably," despite the horrors raging on the sea beyond the galley's metal walls. For most who took part in the Battle of Jutland, however, there were few consolations. As officers read the roll calls, the scale of the human loss became evident. The names of the dead hung in the air like the strains of a dirge. "I read the muster for the forecastle division," Sub-Lieutenant Clifford Caslon on board the *Malaya* recalled. "[I]t was grim business, and I was glad when it was done."

Evidence of the slaughter was inescapable. "Human flesh had got into all sorts of nooks, such as voice pipes, telephones, ventilating shafts and behind bulkheads," Able Seaman Victor Hayward remembered. He and his comrades scrubbed the ship down with carbolic soap to rid it of the pungent smell of rotting flesh, while others went about the macabre task of trying to identify the dead, some of whom had been reduced to nothing but "unrecognisable scraps of humanity."

It fell to the survivors to handle the charred and dismembered remains of their comrades, sewing the bodies into sackcloth with weights in preparation for burial at sea. Despite these measures, many corpses stubbornly bobbed on the surface of the water for several minutes until gradually settling into an upright position and then slipping beneath the waves. "It was an eerie scene as though [the dead] wanted to take one last look at their old ships before they went under," Signalman John Handley remembered.

For the wounded, the end of the battle was only the beginning of their troubles. Stretcher-bearers busied themselves searching for and collecting injured men. Ships' surgeons rushed about treating casualties, working without rest until they collapsed from exhaustion. "[T]hey were so overcome with sheer fatigue, that the last cases were bandaged with the doctors lying down beside the patient, for they could stand no longer," one man observed. The greatest obstacle facing medical teams on board damaged ships was locating suitable space and equipment to tend to those in need. "We had only candle lamps available and they give very little illumination for critical operations," Surgeon Lieutenant Charles Leake recalled. He also noted that the lack of gloves made the surgeons' hands "fearfully sore" due to the use of carbolic acid as an antiseptic.

The worst cases involved burns. "They were so badly burnt that one could do very little to relieve them of the pain . . . injections of morphia seeming to have very little, if any, effect on them," one doctor noted. Even seemingly mild burns could turn fatal very quickly. Duncan Lorimer, a surgeon on the *Malaya*, observed:

> A man will walk into the dressing station, or possibly be carried in, with face and hands . . . not deeply burned, nor disfigured. One would call it a burn of the first degree. Very rapidly, almost as one looks, the face swells up, the looser parts of the skin become enormously swollen, the eyes are invisible through the great swelling of the lids, the lips enormous jelly like masses, in the centre of which, a button-like mouth appears.

Lorimer correctly ascertained that this strange phenomenon occurred when a sailor was burned by superheated cordite—a propellant explosive used to drive a shell or projectile from a gun. The resulting injuries were known as "flash burns." At Jutland, they were often caused by the flash of cordite exploding in a confined space. The detonation was so brief that only exposed flesh was scorched—typically the face, hands, and ankles. It was "quite unlike any burns I have ever seen in civil life," Lorimer remarked. Victims often complained of extreme thirst before falling into acute respiratory distress due to the inhalation of smoke and fumes: "they die and die very rapidly," he wrote. The speed with which this could happen was startling, even to battle-hardened surgeons like Lorimer.

Complicating the situation was the fact that each ship had only a small team of medical personnel on board. In the pandemonium following the battle, the ships' surgeons were soon overwhelmed, and they struggled to treat burn victims quickly and effectively. Frederick Arnold, the wireless telegraphist on board the *Malaya*, described the "grim, weird, and ghoulish sight" of the badly burned who were "almost encased in wrappings of cotton wool and bandages with just slits for their eyes to see through." Some of the surgeons caused further harm in their haste to treat casualties. Alexander MacLean, on board HMS *Lion*, described how he and his colleagues first used picric acid as an antiseptic, which effectively subjected the skin to a tanning process, giving it a hard surface. MacLean soon discovered that the substance also dried out the wound dressings, making them difficult to peel off without removing the raw skin underneath. Only later did MacLean begin applying eucalyptus and olive oil to keep the dressings supple so they could be easily unwrapped.

Unfortunately for many, the effects of picric acid dressings left deep scars. Such damage, caused by well-meaning but overworked medics hastily triaging the wounded, would have to be addressed once the immense undertaking of rebuilding these men's faces began. But the surgeon in charge of the Cambridge Military Hospital's rather more painstaking work would soon find himself just as overwhelmed as his frontline colleagues.

*In the early days of June 1916, naval casualties from the Battle of Jut-*land began pouring into Aldershot by the dozen. Their faces were charred and horribly damaged, and Harold Gillies had never seen anything quite like it. Their injuries had rendered them "hideously repulsive," leaving them "well-nigh incapacitated." He watched in despair as these sailors arrived at the hospital. "How a man can survive such an appalling burn is difficult to imagine," he wrote, "until one has met one of these survivors from fire, and realised the unquenchable optimism which carries them through almost anything."

Gillies quickly set to work accommodating the sudden influx of patients. Gradually, he developed a routine and, more important, a repertoire of techniques for addressing devastating burns. His work during that time was not without precedent. "There is hardly an operation[,] hardly a single flap in use to-day that has not been suggested a hundred years ago," Gillies observed.

The medical treatment of burns was first described in the Ebers papyrus, an Egyptian medical text dating to 1550 B.C.E. The ancient document instructs readers to apply a mixture of cattle dung and black mud to singed skin. In the fifteenth century, the German surgeon Wilhelm Fabricius Hildanus became the first to divide burns into three different degrees of severity. At the same time, debates raged about whether it was best to moisten a burn while treating it or leave it dry and seal the wound. Almost simultaneously, the first documented attempts to excise burned skin were recorded. However, the benefit of removing dead tissue was counteracted by blood loss, poor hygiene, and a lack of antiseptic surgical techniques, which resulted in high rates of infection.

This began to change in the nineteenth century. The American surgeon Thomas Dent Mütter, an early pioneer of plastic surgery, attempted to reconstruct the face of a twenty-eight-year-old woman who had been severely burned as a child when her clothes caught fire. Mütter described how she was "unable to throw her head to the left side or backwards, or to close her mouth for more than a few seconds [at] a time" due to her injuries. Aside from the functional limitations resulting from her burns, she also was severely disfigured. Her right eye was drawn down at an angle, giving her a lopsided appearance.

Mütter proposed a radical solution that was not only risky, but also painful in an era before anesthetics. "To this my patient readily assented," Mütter wrote. Using a series of rotation and advancement flaps, he was able to transform her appearance: "the patient [was] so much altered that persons who saw her before the operation, scarcely recognized her as the same individual." Mütter went on to publish extensively on his reconstructive work on burn victims—most of whom he treated years or decades after their initial injuries.

Although published work on burns by early practitioners would have been known to Gillies, its instructional value to him was limited. Gillies wrote that his reconstructive work at Aldershot was "original in that all of it has had to be built up again *de novo*." He quickly discovered the impracticability of previous methods, which he believed must have been "put forward on the study of one case only, or even on purely theoretical grounds." For instance, while Mütter was often able to alter a patient's appearance, he was not always able to restore function. "There are cases in which we must be content with this while the loss of the function is an evil for which there is no remedy," Mütter lamented.

In contrast, Gillies was just as concerned with function as he was with form, and he knew the two were intrinsically linked. He warned about the dangers of a "presentable appearance," which, he argued, could be "the mask of a skeleton of surgical inefficiency." He worked from the inside out, reconstructing internal membranes first, followed by supporting structures such as bone or cartilage, and, last, the skin. By doing this, he was able to achieve results that were both aesthetically pleasing and functional. "In planning the restoration, *function* is the first consideration," he wrote, "and it is indeed fortunate that the best cosmetic results are, as a rule, only to be obtained where function has been restored."

Gillies settled into a rhythm, working diligently and unfalteringly on his charges. One day, he invited Sir William Arbuthnot Lane to visit the unit. Gillies was keen to showcase one of his prize patients, on whom he had performed a procedure to correct a contorted lip. On the day of the visit, Gillies—usually a paragon of cool confidence—couldn't help but feel anxious in the presence of the senior surgeon who had helped him establish the specialty unit at Aldershot.

Lane was famous within the medical community for his exacting nature, especially his fanatical commitment to asepsis in the operating room. To minimize the risk of infection, he developed a "no touch" technique that involved using surgical instruments with elongated handles to ensure that even a gloved hand would not come in contact with any part of a wound. "[W]ith long in-struments he could fashion a chicken bone graft and introduce it without once touching patient or graft with his hands," Gillies marveled. Lane's dexterity had even inspired someone to draw a cartoon of him deftly performing surgery through a hole in the dome of his operating theater using absurdly long instruments.

On arriving, Lane was shown around the unit by Gillies and Nurse Black. Tonks was also present, trailing behind the group with his sketch pad and pencil. When they reached the patient in ques-tion, Gillies directed Black to remove the man's bandages. After a brief pause, Lane leaned forward and gently pressed an instrument into the newly sutured lip. "To my horror a great drop of pus oozed out," Gillies recalled with commendable honesty. He was never one to abrogate responsibility for his mistakes.

And there were plenty of mistakes in those early days. Recon-structed noses shriveled up due to the lack of a mucous membrane. Skin grafts failed to take. Flaps became infected. "Then I had to confess to the boys that I had made a mess of it and that we would have to start again," Gillies said. "It was not easy."

The visit may not have gone as Gillies had hoped, but Lane was nonetheless impressed by the work being done there. Before leav-ing, he told Gillies that he would be allocating a further two hun-dred beds to him, in anticipation of a major new offensive. What Gillies had experienced in dealing with the human cost of the Battle of Jutland would pale in comparison to what was about to happen on a battlefield that would become synonymous with the Great War itself.

≯ 6 ≮

THE MIRRORLESS WARD

The early-morning sun was already powerful as British troops readied themselves for the Somme offensive on July 1, 1916. It was certain to be a sweltering day, especially for men bogged down by heavy uniforms, helmets, and weapons, who would have to make their way across a battlefield devoid of any cover or shade.

Private R. W. D. Seymour, known as "Big Bob" to his friends, was peering out over the wall of his trench. Although apprehensive, many of his comrades were in high spirits. They felt confident that the weeklong artillery bombardment of the German lines had weakened the enemy and that they would encounter little resistance once the "big push" began. "We were informed by all officers from the colonel downwards that . . . there would be very few Germans left to show fight," recalled Lance Corporal Sidney Appleyard of Queen Victoria's Rifles. They were confident that the attack would be over quickly and that their actions that day would prove a significant step toward ending a god-awful war. Little did they know that more than half of the artillery shells had failed to detonate, leaving most of the Germans' fortified bunkers, deep dugouts, concrete strongpoints, and wire barricades largely intact.

Suddenly, an unearthly explosion convulsed the landscape south of the village of La Boisselle. Staggering quantities of soil and debris flew into the sky. When the dust had settled, a crater 330 feet wide and 90 feet deep was left behind. In coming days, the blast that had created it would be deemed louder than any man-made noise previously recorded; there were even reports that it had been heard in London, 190 miles away. The charge laid in the Lochnagar mine, named after the British trench from which it was dug in secret by the Royal Engineers, was one of nineteen set in tunnels burrowed under the German lines. They were designed to help the British infantry as they advanced toward the enemy that morning.

Two minutes after the explosion, that advance began. The air was pierced by a shrill chorus of officers' whistles. Private Seymour and one hundred thousand of his fellow troops swarmed out of the trenches like ants out of a nest. As Seymour and his comrades pushed forward, men began dropping under a deluge of explosives, shrapnel, and gunfire. They fell so quickly that it looked as though they had been ordered to lie down. One German machine gunner recalled the ease with which he and his comrades were able to shoot troops advancing across the open terrain from their enclosed positions. "When we started to fire we just had to load and reload. They went down in the hundreds. We didn't have to aim, we just fired into them," he wrote. Visibility was limited due to the clouds of dust being kicked up by exploding shells. One eyewitness described it as "a veritable inferno." Unbeknownst to anyone at the time, the darkest day in the history of the British Army had just begun.

Seymour had only advanced a short distance before heavy fire forced him to take shelter in a shell hole. From there, he spotted a German officer in the distance, frantically signaling to his men. Big Bob saw an opportunity. He moved from his place of safety, took aim, and fired his Lee-Enfield rifle at the officer. As the man fell, Seymour dropped to his knees in triumph. Just then a shell exploded, peppering his face with shrapnel and shearing off half of his nose. The force of the impact spun Seymour around, at which point he was shot five times in the back by a machine gunner. He lay gravely injured where he fell while the fight continued to rage around him.

The battlefield was soon littered with the dead and dying. "There were men everywhere, heaps of men, not one or two men, but heaps of men everywhere, all dead," one soldier recalled. Donald Murray of the King's Own Yorkshire Light Infantry was horrified by the sight of "men on the barbed wire with their bowels hanging down, shrieking." He described it as a hell made of fire and smoke and stink. George Rudge—who was only seventeen at the time— took in the scene in bewilderment. "It seemed to me that everyone around me had been killed or wounded for I was the only one of my regiment I could see," he observed.

Of the 100,000 British soldiers who took part in the advance, 19,240 men were killed and a further 38,230 wounded—most of them gravely. Never before or since has a single army suffered such losses on a single day in a single battle. This was in stark contrast to the German army, which suffered around 6,000 casualties on the first day of the Battle of the Somme. The ground the British took could be measured in yards rather than miles. Both sides were trapped in a small area with an enormous amount of firepower, and there would be no end to it for another 140 days. When reports of the carnage reached the shores of Britain, the names of the dead from the first day of the battle occupied not columns but entire pages in the newspapers.

Within hours of the start of the Somme offensive, casualty clearing stations near the Western Front were overrun by wounded men in urgent need of medical assistance, among them Private "Big Bob"

Seymour. Worsening the situation was a planning failure: there were not enough trains to transfer patients to base hospitals that first day. As a result, tens of thousands of soldiers lay slumped in entranceways, corridors, recreation rooms, and dining halls. They spilled out onto the grounds in front of the facilities in such numbers that, by the end of the day, one nurse remarked that not a single blade of grass remained visible. Jack Brown, a medical orderly, vividly recalled the bloody chaos:

> We had so many wounded coming in we all lost track of time . . . It was my job to line up the cases needing surgery on stretchers outside the operating theatre . . . It was up to me to decide who should have surgery and who was too far gone . . . it was terrible, I can never forget it, no one should have to do that . . . but the surgeon couldn't, he was too busy operating on them. I knew that he had to sort them out quickly otherwise they'd all die.

Soldiers marked with bright red stripes on their uniforms were rushed into rudimentary operating rooms for emergency treatment to stop them from bleeding out. They were carried past piles of amputated limbs. "Every so often the surgeon told me to clear [the limbs] out, and I had to find somewhere to burn them," Brown remembered. "I felt sorry for those lads outside who were waiting to go in and who could see all this going on."

Philip Gibbs, who served as an official British reporter during the First World War, noted that the "hospital huts and tents were growing like mushrooms in the night." Every conceivable mode of transport was employed to evacuate the wounded and make way for new casualties. "They were cleared out of the way so that all the wards might be empty for a new population of broken men, in enormous numbers," Gibbs wrote. "There was a sinister suggestion in the solitude that was being made for a multitude that was coming."

Within days, injured soldiers began arriving at the Cambridge Military Hospital in great numbers. Some had Gillies's handwritten labels pinned to their tattered uniforms; others were wearing official labels from the War Office. Gillies would not allow himself to feel daunted by the scale of the situation. He watched a "grotesque procession" of wounded men disembarking from hospital trains and making its way onto the ward. As he did, he thought to himself, "Let us roll up our sleeves, for the work really begins now."

There would be no respite for weeks. "There were wounds far worse than anything we had met before," Gillies wrote in a letter to his friend Dr. Lyndon Peer. "My days and nights were filled with a steady flow of injuries." The staff began to feel the strain. "In all my nursing experience those months at Aldershot . . . were, I think, the saddest," Catherine Black wrote. Maintaining morale was a constant struggle and required infinite patience and compassion on the part of the doctors and nurses. The psychological toll on the wounded could be overwhelming. "Hardest of all was the task of trying to rekindle the desire to live in men condemned to life week after week smothered in dressings and bandages, unable to talk, unable to taste, unable even to sleep without opiates because of the agony of lacerated nerves, and all the while knowing themselves to be appallingly disfigured."

Private "Big Bob" Seymour was eventually rescued and sent to a base hospital, where he spent time recovering from the multiple gunshot wounds to his back. Afterward, he was transferred to Aldershot so that the damage to his nose from the shell explosion could be addressed. Until that point, Gillies had not had many opportunities to reconstruct noses, since most of the casualties on his wards had suffered damage to their lower jaw and face. It took the devastating carnage of the Battle of the Somme to extend the purview of his work. Seymour would be one of the first of many "nose jobs" Gillies would perform during the war.

Rhinoplasty, an operation to alter the appearance of the nose, is one of the oldest recorded surgical procedures in history. The Indian surgeon Sushruta, who lived more than two thousand years ago, is credited with devising a method of nasal reconstruction, a version of which is still used today. This involved cutting a flap of skin from either the forehead or the cheek and attaching the free end to the bridge of the nose. Two small reeds were inserted into the nostrils to facilitate breathing while the nose healed and the swelling subsided. After the free end attached itself to the new site, the flap was then severed from the forehead or cheek and sewn over the damaged area, providing a serviceable replacement for the lost nose.

A similar technique appeared in Europe shortly before the Renaissance. In 1432, a surgeon named Gustavo Branca obtained a license to open a specialist shop in Catania, Sicily, where he used skin flaps from the cheek and forehead to reconstruct noses. A few years later, his son Antonio improved upon this method by opting for a more discreet donor site: the arm.

This new technique involved partially cutting a flap of skin from the upper arm, reshaping it into a nose, and then attaching it to the damaged nasal cavity. The arm would then be held in place against the head using bandages for as many as forty days. Afterward, Branca severed the new "nose" from the arm and began reshaping and contouring the remaining skin.

This method, which eliminated the need to further mark the face by taking skin flaps from the forehead or cheek, was popularized in the sixteenth century by an Italian surgeon named Gaspare Tagliacozzi, who improved upon the technique. He boasted that his clients received noses "so resembling nature's pattern, so perfect in every respect that it was their considered opinion that they liked these better than the original ones which they had received from nature." Tagliacozzi's work was motivated in part by an increase in nasal injuries inflicted by rapiers, the duelist's new weapon of choice. But many people sought a surgical solution due to the growing association of nasal disfigurement with syphilis, a disease that had made its first appearance in Europe around this time.

Those who contracted syphilis often developed "saddle nose," which occurs when the bridge of the nose caves into the face. As a consequence, nasal disfigurement was seen as a sign of moral failing in its victims, regardless of the actual cause. This stigma persisted for hundreds of years. In 1705, the satirical writer Edward Ward warned that the "French Pox" (i.e., syphilis) would "lead [sufferers] by the Nose into publick Shame and Derision." Indeed, the fear of nasal deformity was so great that noses were often injured purposely as a form of punishment, particularly for sexual transgressions like prostitution or adultery.

Given this stigma, it's unsurprising that people turned to Tagliacozzi, despite the high risk of infection and further disfigurement. He believed that the surgeon's task was to "restore, repair, and make whole those parts of the face which nature has given but which fortune has taken away, not so much that they might delight the eye but that they may buoy the spirits and help the mind of the afflicted." In 1597, Tagliacozzi published *De Curtorum Chirurgia per Insitionem* (*On the Surgery of Mutilation by Grafting*)—the first book to deal exclusively with reconstructive surgery. Large sections of it are devoted to rhinoplasty.

Strange tales abound of noses being completely severed and reattached. In the eighteenth century, the French surgeon René-Jacques Croissant de Garengeot told of a soldier whose nose had been partially severed during a fight: "some wine was warmed to cleanse the wound and his face which was covered in blood." Garengeot reported that the nose was then placed in the wine "to warm it a little," after which it was "successfully adjusted to its natural position" and kept in place using plaster and tape. Something similar supposedly occurred in the early nineteenth century when "a Spaniard [named] Andreas Gutiero struggled with a soldier who cut off his nose and there it fell down in the sand." A surgeon, who just happened to be present, urinated on it before reattaching it to the unfortunate man's face. In both cases, the nose was allegedly grafted back onto the patient's face successfully—though there is reason to doubt the veracity of these accounts.

Not everyone viewed these surgical developments positively. The historian Sander Gilman points out that the restoration of a diseased nose enabled its owner to pass as healthy and was a manifestation of the power of the surgeon to remake man in his own image. In an era when people believed—quite literally—that illness was a punishment from God and that physical imperfections mirrored the status of the soul, the idea that someone could mask a deformity by undergoing reconstructive surgery was seen by some to be immoral, if not downright dangerous. This is one of many reasons rhinoplasty faded into obscurity after Tagliacozzi's death in 1599. It only reappeared in the early decades of the nineteenth century, when a British surgeon named Joseph Carpue resurrected and disseminated Sushruta's "Indian method" in his book *An Account of Two Successful Operations for Restoring a Lost Nose* (1816). It was at this time that rhinoplasty enjoyed its own renaissance in surgery.

Even though rhinoplasty had been around for centuries, existing methods were ineffective in addressing the severity and variety of nasal damage inflicted during World War I. This was especially true if the bridge of the nose or the cartilage had been destroyed, since most older techniques only addressed soft tissue reconstruction through the use of skin flaps. But surgeons discovered that even cases involving flaps could be problematic.

One patient who found himself in Gillies's care had first undergone reconstructive surgery in Birmingham. The rotation flap used to reconstruct the nose had not been supported by a cartilage bridge, so the entire structure had collapsed inward. Additionally, the surgeon had taken the flap from the patient's forehead but, in doing so, accidentally transplanted part of his scalp onto the new tip. By the time the man reached Gillies, "a sizable tuft of hair" was protruding from his new nose. It took twenty-one operations and nearly five years to correct and improve the defect.

Soon after Private Seymour's own arrival at the Cambridge Military Hospital, Gillies handed him an album of different nose types for his perusal. After some consideration, Seymour decided on a Roman nose with a prominent bridge. Over the course of two operations, Gillies rebuilt Seymour's nose.

First, he harvested a piece of cartilage from the patient. To establish a blood supply, Gillies wrapped it in a blood-vessel-rich flap of tissue that was folded downward from the forehead. Afterward, this was covered in what Gillies termed the "Bishop's Mitre" flap—so named because the skin used to cover the site was cut in the shape of a truncated kite resembling a bishop's hat. Two months later, Gillies was able to advance the flap downward, since the cartilage had produced a satisfactory support for the new tip.

It was an imperfect first attempt, giving Seymour more of a swollen boxer's nose than the aquiline profile of a senatorial Roman. But Seymour was so pleased with the outcome that he agreed to become the surgeon's private secretary after his recovery, a position he held for the next thirty-five years.

Repairing nasal injuries became a commonplace procedure for Gillies. But one case stood out from all the others. So successful was it that Gillies would later note its significance to the improvement of nasal reconstruction techniques.

William Spreckley—the eldest son of a lace-maker—was in Germany learning his trade when the war broke out. As he made his way back to Britain, he was stopped by authorities. On account of his fluent German, the men who detained Spreckley mistook him for a German citizen and thought he was attempting to flee the country to avoid fighting. Eventually, Spreckley made it back home to England, where he enlisted in the army and was sent to Belgium. There, he rose to the rank of lieutenant before sustaining the facial injury that would cut short his military service.

When Spreckley arrived at the Cambridge Military Hospital in January 1917, a large crater occupied the middle of his face where his nose had once been. Gillies met his new patient with the same quiet confidence with which he met all those who came under his care. "Don't worry, sonny," he said, despite being only a few years older than Spreckley. "[Y]ou'll be alright and have as good as face as most of us before we're finished with you."

Gillies wasted no time getting to work. Time was of the essence, given the depth of the gash across Spreckley's face. First, Gillies applied skin grafts to the raw areas of the wound to ensure that the airways were protected. He then took a piece of cartilage from below one of Spreckley's ribs and shaped it like an arrowhead so that it would eventually give lateral support to the wings of the nose that form the nostrils. Gillies implanted this into Spreckley's forehead near the hairline, where it remained for six months. Next, he created a skin-graft inlay to supply a future nasal lining and placed this below the cartilage. After establishing a viable blood supply, Gillies swung the cartilage and skin-graft inlay downward to construct the bridge of Spreckley's nose. He then covered the bridge with a skin flap taken from the soldier's forehead.

Initially, Spreckley's new nose was gargantuan—three times the size it should be. Where once there had been nothing but a giant pit in the middle of the lieutenant's face, there was now a bulbous mass of skin and tissue. Gillies likened it to an anteater's snout: "all my colleagues roared with laughter . . ." Discouraged by the result, Gillies lost faith in the complex technique, vowing never to repeat it. But soon enough, the swelling began to subside, and he was able to remove the excess fibrous tissue around the site. The result was encouraging, as the semblance of a nose began to appear. Gillies wrote, "[H]asty judgment leads often to the discard of the principle the soundness of which may later be proved."

As Spreckley's wounds healed and his nose settled into place, he became one of Gillies's star cases. "Look at Spreckley today," Gillies joked later in life. "He and his nose went back to the Army in 1939 and served together until [he left the military in] 1950." It was a happy ending for both the surgeon and his patient.

>─■─◄

With limited resources at his disposal, Gillies was under intense pressure to figure out how best to reconstruct the faces of the countless men arriving on his doorstep each day. It was his patients' faith in him that boosted his morale during that awful time. "[W]ithout it I was lost," he wrote.

Gillies had to rely on his imagination to visualize complex surgical procedures and could often be found making quick sketches on the backs of envelopes when ideas occurred to him. The sheer number of patients arriving at the Cambridge Military Hospital also afforded him the opportunity to experiment with different techniques. Reflecting on those days at Aldershot, he remembered, "All the time, we were fumbling towards new methods and new results . . . My staff and I felt that we were on trial."

The tomblike quiet blanketing Gillies's domain heightened the feeling of being on trial. Nurse Black referred to the unit as "that silent ward where only one in every ten patients could mumble a few words from the shattered jaws." Worse than the deathly quiet was the occasional sound of a patient screaming in agony. Gillies wondered if some of his patients hadn't died of a broken spirit during the dark aftermath of the Somme offensive.

When he first began working at the Cambridge Military Hospital, Gillies banned mirrors on his wards. The ban not only protected new arrivals from the shock of seeing their injuries for the first time, but it also protected those in the midst of lengthy reconstructive surgeries from seeing their faces before the work could be completed. Captain J. G. H. Holtzapffel remembered his reaction upon seeing his nose shortly after the initial operation: "When I first got a chance to examine myself in the looking glass I got a bit of a shock, for my beautiful new nose looked more like a short piece of cucumber slapped on my face." Gillies understood the effect this could have on a man's willingness to continue with reconstructive work. "If our plastic plans went wrong," he explained, "a patient without great moral fibre would drift into a state almost of delinquency." Only those blinded in combat remained in good spirits while their faces were rebuilt, Gillies observed.

Preventing men from catching sight of their own reflections wasn't always easy. To protect his identity, "Corporal X" was the designation Nurse Black used in her notes for a soldier who arrived at Aldershot shortly after the Somme offensive began. Like so many men arriving at the Cambridge Military Hospital, he was still caked in mud from the trenches, half his face blown apart by shrapnel.

For the first several days, Corporal X slipped in and out of consciousness as his wounds festered and became septic. Black remarked that no one, not even Gillies himself, believed that the young man would pull through. But thanks in part to her constant ministrations, which included around-the-clock feedings, the soldier rallied. Although the injuries to his face were severe, Corporal X had not lost his ability to speak. It wasn't long before he regaled everyone on the ward with stories about his fiancée, Molly.

Corporal X had been in love with Molly ever since he had met her at a dance class as a child. When he turned eighteen, he went to law school and, after receiving his degree, returned home to set up his own practice. He worked hard over the next few years to build up his clientele before asking Molly for her hand in marriage. Even though his business was thriving, he was afraid she would reject him, as her parents were wealthy landowners and were openly opposed to the union. But it turned out that Molly was just as smitten with him as he was with her, and she happily accepted his proposal in spite of her parents' objections.

When the war broke out, he volunteered for service immediately—a decision frowned upon by Molly's parents, who felt he should wait for a commission as an officer. Molly supported his decision and wrote him letters each week, sharing her dreams and plans for them when he returned. Her words cheered him during his darkest moments—the darkest of all being his protracted recovery from having scalding hot shards of metal tear through his face.

"I don't want her to come until I get some of these beastly bandages off," he told Black one day as she was tending to his wounds. "It would scare her to death to see me lying here looking like a mummy." Corporal X, who had not seen himself since before he was injured, clung to the hope that his disfigurement would be mild.

On the day that his bandages were finally removed, his mother happened to be visiting. "She went very white," Black recalled. "I thought for a moment that she was going to faint, but not the slightest expression of face or voice betrayed her." As Black delicately unwound the wrappings, the corporal's mother continued making small talk, even though the man before her bore little resemblance to the handsome son she remembered. Later that evening, the young man called for Black, asking that she put screens around his bed. As she did so, the glint of a shaving glass hanging in his locker caught her eye. To her dismay, she realized that Corporal X had seen his face. "Every nurse learns that there are moments when it is better to leave a patient alone because sympathy would only make things worse."

Corporal X sank into despondency. The future he had imagined for himself seemed to die with that glimpse of his reflection. He had internalized his society's sense of repulsion at the sight of a disfigured face and had turned it on himself. He no longer felt worthy of love due to his altered appearance. Indeed, the ban on mirrors likely reinforced in him a feeling that his was a face not worth gazing upon. Black reckoned "he must have fought out his battle in the night." The next morning, he asked her to post a letter addressed to Molly. After she had done so, Black returned to the ward and said to the young man, "You're well enough to see her any time now. Why not let her come down?"

With sorrow in his voice, Corporal X answered quietly, "She will never come now." He then told Black that he had lied to Molly in his letter, informing her that he had met a woman in Paris and had realized that their engagement was a mistake. "It wouldn't be fair to let a girl like Molly be tied to a miserable wreck like me," he said to Black. "I'm not going to let her sacrifice herself out of pity. This way she will never know."

Black was dismayed by this turn of events. She lamented that "Gillies had done everything that was humanly possible, but he could not work miracles." The same standards of beauty that deemed Gillies's work necessary also led to some patients being seen as "failures"—even by themselves—when surgery could not alter their appearance in ways that met those standards. In those early days, Gillies was still learning—and largely inventing—his craft, and doing so on the most challenging cases imaginable. A high proportion of sad and unfavorable outcomes was inevitable.

Arbuthnot Lane reckoned that "[n]othing was more painful than the sense of loneliness [in these men]." In the case of Corporal X, he was not wrong. When at last the young man was discharged from the Cambridge Military Hospital, he went home and chose to live the quiet life of a recluse.

Broken faces frequently led to broken hearts during the war. Shortly after the Somme offensive began, Private Walter Ashworth of the 18th West Yorkshire Regiment found himself at the Cambridge Military Hospital. He was one of only a handful of soldiers from his unit who survived the first day of battle. Faced with intense fire from the enemy, most of the men hadn't even made it to the front line of their own trench before falling dead on the turf of the battle-field. Ashworth—who lost half of his face after a bullet ripped into his cheek and shattered a large section of jaw—had tumbled into a water-filled crater and lay there for three days, until someone noticed he was still alive and dragged him to safety.

He was quickly transported back to Britain, where he was ad-mitted to the Cambridge Military Hospital on July 5, 1916. Soon after his arrival, Henry Tonks captured the soldier's likeness as nurses irrigated the terrible gash in his face. In the pastel drawing, Ashworth is slouched over a kidney basin, which has been placed under his chin to catch the water, blood, and mucus pouring from his wound. His piercing blue eyes stare off into the distance, while a wisp of hair falls over his brow. It is the portrait of a man who is only beginning to grasp the horror of his situation.

Ashworth's case was especially challenging due to the amount of bone and tissue loss he had suffered. "Unfortunately the missiles [from the battlefield] were not merely lacerating and fracturing, but were tearing away large facial hunks, which meant that pieces were actually missing from the puzzle," Gillies wrote. By this time, he understood that simply pulling the adjacent flaps of skin together would not yield good results unless the underlying structure was repaired first. Complicating matters was the fact that the fractured jawbone had to be realigned and immobilized before surgery could even be attempted. Where there was no bone or tissue loss, this could be achieved through the use of a dental splint that held the bone fragments together while they healed. The dentist could either use a cap splint, which fit over the teeth, or an external splint that was affixed to the outside of the face when a patient had suffered extensive tooth loss due to injury or decay.

If "facial hunks" were missing, however, the realignment was only successful as long as the splint remained in place. This meant that the dental apparatus had to be worn at all times while Gillies created a bone or cartilage graft to reinforce the underlying structure. It could take anywhere from three to twelve months for a graft to unite, during which time the patient would not be able to eat solid food due to the immobilization of the jaw. The solution was a liquid diet, but even this became tricky if the patient had also suffered damage to his hard palate (the roof of the mouth), since the liquid could pass up into the nose. Making matters worse was the fact that not being able to move one's jaw for long stretches of time could cause the temporomandibular joint that hinges the jaw to the skull to lock. Nothing was easy or straightforward. And no reconstruction was achievable without the help of competent dental surgeons.

Ashworth underwent three painful operations to reconstruct his shattered face. A rare surgical diagram from this procedure survives. It shows how Gillies closed the wound by suturing together flaps of skin and tissue from the cheek and jaw. Gillies later wrote that it had been necessary to sacrifice some of the length of the lips to close the gaping hole in Ashworth's cheek, and, as a result, the patient was left with a "whimsical, one-sided expression that . . . was not entirely unpleasant."

Unfortunately, Ashworth's fiancée felt differently. After learning of his facial disfigurement, she broke off their engagement. Louise Grime, a friend of the fiancée, heard about this painful snub. Moved by the heartbreaking situation, she began writing to

Ashworth at Aldershot. The pair exchanged several letters before Grime plucked up the courage to ask if she could visit the young soldier in the hospital. He enthusiastically agreed. Soon, the two fell in love.

Once Ashworth was discharged from the army, he traveled back to his hometown of Bradford, where he had worked as a tailor before the war. Unlike William Young—the recipient of the Victoria Cross, who died from complications of anesthesia under Gillies's care—Ashworth was not given a hero's welcome. No marching bands or parades greeted him on his arrival, and his former fiancée wasn't the only one who reacted negatively to his altered appearance. When he returned to his old job, his boss insisted he perform menial tasks at the back of the shop so that customers would not be frightened by his appearance. The demotion upset him so much that he gave notice and left. Wounds were not inflicted only on the battlefield.

Ashworth's relationship with Grime continued to blossom despite these personal setbacks. He eventually proposed, and the two were married. After the war, the couple moved halfway across the world to Australia in search of a new beginning. Many years later, Ashworth bumped into Gillies, who was there on a teaching visit. The surgeon wondered aloud if he might have another attempt at his former patient's face. According to his granddaughter, Ashworth thanked him but declined. Perhaps he had made peace with the face that Gillies had given him during a time when he had thought all hope of a normal existence was lost.

⋊ 7 ⋉

TIN NOSES AND STEEL HEARTS

s Harold Gillies toiled away at the Cambridge Military
Hospital, elsewhere a British artist named Francis Derwent Wood was developing his own idea of how to help
disfigured soldiers. Born in 1871 of an American father and a
British mother, Wood trained at various art institutes around the
world, cultivating a talent for sculpture that he had first evinced as
a child. As a student of the famous sculptors Édouard Lantéri and
Sir Thomas Brock, Wood became a well-respected artist in his
own right, exhibiting his work at the Royal Academy every year
from 1895 until his death in 1926.

Like Tonks, Wood was too old for active duty when the war
broke out. At the age of forty-four, he enlisted as a private in the
Royal Army Medical Corps and was assigned to the 3rd London
General Hospital in Wandsworth, where he worked as an orderly
designing splints. While there, Wood was deeply affected by the
reaction of visitors to patients with facial injuries. He realized
his artistic abilities might be useful and so began constructing
masks for these men, many of whom had suffered extensive tissue

loss and had already undergone multiple operations. "My work begins where the work of the surgeon is completed," said Wood.

The concept of facial prostheses has a long history. In 1566, the famous astronomer Tycho Brahe had part of his nose sliced off during a duel, and he subsequently wore a replacement that he attached using a glutinous substance that he carried with him in a small metal box. According to legend, the nose was made from silver, but when his body was exhumed in 2010, scientists performed a chemical analysis of the bone around his nasal cavity and discovered the prosthetic had actually been made of brass.

Disease also played a role in the proliferation of such appliances. Many people whose appearances had been ravaged by syphilis turned to nasal prosthetics to disguise the unsightly signs of infection. In the 1860s, American inventors John Wesley Hyatt and his brother Isaiah discovered that nitrated cellulose mixed with camphor produced a substance that could be shaped or molded. Celluloid—the first plastic—quickly became the material of choice when producing artificial noses. Unfortunately, it was highly flammable. Given the prevalence of smoking in the late nineteenth century, stories abound of celluloid noses turning brown or even catching fire.

War had given rise to opportunities for both surgical and artistic innovation in earlier centuries as well. A book by the sixteenth-century military surgeon Ambroise Paré contains illustrations of enameled eyes, ears, and noses made from silver and gold, which were used by wounded soldiers. When Private Alphonse Louis—a twenty-two-year-old French artilleryman—was struck by a seven-pound piece of shrapnel in the Siege of Antwerp in 1832, he lost a large portion of his lower jaw. Louis was transported to a field hospital, where a surgeon attempted to close the wound by drawing the remaining soft tissue together. His prognosis was dire. Louis's tongue swelled to four times its normal size, which prevented him from eating and drinking effectively. During his recovery, Louis subsisted on a mixture of thin broth and lemonade mixed with wine, which was administered by a curved spoon placed at the back of his tongue.

Once his condition stabilized, he came under the care of one Dr. Forjet, a surgeon-major to the Army of the North. Forjet first created a plaster cast of Louis's damaged face. He then enlisted the help of a master craftsman, who engineered a mask made from silver. The mask, which weighed three pounds, had articulated components that allowed the mouth to open so Louis could eat. Attached to the interior of the mask was a drainage chamber that collected saliva. The same attention and care lavished upon the mechanics of the mask were extended to the exterior, which was painted to match Louis's complexion and furnished with a mustache made of real hair. Louis became famous in medical circles and beyond as "the gunner with the silver mask."

In spite of these successful cases, it wasn't until the First World War that facial prosthetics were produced in large numbers. And much of the work began with Wood, who established the Masks for Facial Disfigurement Department at the 3rd London General Hospital in March 1916—around the same time that Gillies set up his own plastic surgery unit at Aldershot. It wasn't long before people began referring to his department as "The Tin Nose Shop." As it turned out, this was something of a misnomer.

Wood's creations began with a plaster-of-Paris cast of the patient's face. Then, referring to prewar photographs of his subjects, he used clay to fill in gaps on the cast caused by tissue and bone loss. After this process was complete, Wood recast the "new face" and electroplated it with copper. This formed a thin skin of metal, tailor-made to fit the patient's underlying facial structure. He then fastened onto it any necessary accessories—such as spectacles or glass eyes—before giving it a coating of silver. Next, he hand-painted the mask to match the complexion of the patient, before adding eyebrows and eyelashes made from tinted metallic foil that were soldered onto the prosthetic.

Wood's new metallic masks—which were lighter in weight than the vulcanite rubber prosthetics used previously—were custom designed to restore to the wearer his prewar appearance. He did not try to improve upon the man's physical appearance. "As they were in life so I try to reproduce them, beautiful or ugly; the one desideratum is to make them natural," he wrote. (In contrast, Gillies was comfortable creating entirely new looks for his patients, as he did with Private "Big Bob" Seymour, and he often greeted a soldier by asking him what sort of face he wanted.)

Unlike artificial limbs, the production of masks could never be standardized due to the diversity of the injuries and the artistic skill required to restore the unique appearances of individuals. Like reconstructive surgery, each mask was highly bespoke. The production was tedious and time-consuming, with a prosthesis taking approximately a month to construct. Additionally, a mask required frequent adjustments, since healing or the formation of scar tissue might alter the contours of the face over time. If the changes to the underlying tissue structure were too great, Wood would have to begin all over again.

Nevertheless, Wood's masks were soon being hailed as "magical." A journalist at *The Times* wrote that the artist was able to "rob war of its ultimate horror" with his creations. In an article for *The Lancet*, Wood described his painstaking efforts to re-create a man's face as closely as possible to its appearance before he was wounded. He likened the psychological impact of the masks to what patients experienced when undergoing successful reconstructive surgery. "The patient acquires his old self-respect, self-assurance, self-reliance, and, discarding his induced despondency, takes once more to a pride in his personal appearance," he claimed. His work inspired other artists to take up the cause. Most notable among them was American sculptor Anna Coleman Ladd.

Born Anna Watts in 1878 to wealthy American expatriates,

she was raised in Paris, where she received private instruction in modern languages and the arts. In her early twenties, she moved to Rome and studied sculpture. It wasn't long before she met and married Harvard-trained physician Maynard Ladd. The couple relocated to Boston, where she carved out a role as a prominent society artist, creating decorative fountains and portrait busts for the city's elite. When the First World War broke out, Ladd's husband traveled to France to serve in the Children's Bureau of the American Red Cross. There, he established a series of hospitals and relief stations for women and children affected by the war.

Ladd was not the type of woman who was content to sit on the sidelines. Inspired by Wood's work, she began lobbying the American Red Cross to establish a studio in France. During that time, she also corresponded with Wood, who shared descriptions of his techniques with the eager artist. In 1917, Ladd traveled to Paris, where she first spent time at the Val-de-Grâce military hospital observing the work of Hippolyte Morestin—the cantankerous surgeon who had once shut Harold Gillies out of his operating room. In November of that same year, her campaign bore fruit, and Ladd was able to open the Studio for Portrait Masks under the auspices of the American Red Cross.

Situated in the city's Latin Quarter, Ladd's studio was spacious and bright, overlooking an enclosed courtyard overgrown with ivy and crowded with classical statues. She filled the space with fresh bouquets of flowers to create a cheerful and welcoming atmosphere for her clients. After climbing five flights of stairs, visitors entering the studio were greeted with casts of other disfigured men hanging on the wall. On any given day, there were usually half a dozen French soldiers lounging around, smoking and playing dominoes while they waited to be transformed by Ladd's artistic wizardry. Each Tuesday, Ladd arranged tea parties in her studio so that those with facial prosthetics could demonstrate to those yet to receive them that it was possible to restore "normalcy" to their appearance. The disfigured men—known as *les gueules cassées*, or "the broken faces" in France—even shared gifts around a tree at Christmas. "They were never treated as though anything were the matter with them," Ladd recalled. "We laughed with them and helped them to forget. That is what they longed for and deeply appreciated."

Like Wood, Ladd stepped in when surgery failed—a frequent occurrence in the chaos of war, since soldiers were often sent first to surgeons with limited experience in plastic work. Ellen La Motte, the American nurse and journalist who volunteered to the front before the United States entered the war, encountered a French soldier who had been so severely injured that he had to have all four limbs removed. Additionally, his face had a "hideous flabby heap [for a nose] fashioned by unique skill out of the flesh of his breast." His surgeons had done little to correct his disfigured mouth. "[A]ll the front teeth were gone," she wrote, "and in his pocket there was an address from which artificial eyes might be purchased." After the soldier had been sent home, his depression worsened. He had been told that he was a medical miracle, and yet "[he] kept sobbing, kept weeping out of his sightless eyes, kept jerking his four stumps in supplication, kept begging in agony: 'Kill me, Papa!'"

Some soldiers gave up hope, but others turned to Ladd after their surgeons had set down their knives. "One man who came to us had been wounded two and a half years before and had never been home," Ladd wrote about a client. "He did not want his mother to see how badly he looked. Of all his face there was only one eye left, and after 50 operations . . . he came to us." Ladd relied on her anatomical knowledge and artistic intuition to design a mask that provided a reasonable facsimile of the soldier's former face. Like Wood, she worked backward from prewar photographs to achieve a startling, lifelike result. "In order that the [mask] might be as perfect as was humanly possible, the soldiers' faces were studied day after day, in repose, in animation and all photographs possessed by them were brought into service to help in the work," she explained. "[T]hen, working from photographs, or verbal descriptions, I would build up the missing or ruined features, with their habitual or natural expression."

Ladd's workload was heavy, so she enlisted the help of four people—Diana Blair, from Harvard Medical School, and the sculptors Jane Poupelet, Louise Brent, and Robert Vlerick. Together, they made ninety-seven masks during the eleven months Ladd spent in France. Each one sold for eighteen dollars, a modest price given the amount of work required to make each one. A journalist visiting Ladd spied several of these exquisitely crafted masks on a nearby table. So realistic were they that she remarked, "They looked for all the world like human noses and chins laid out for a cannibal's supper."

Although Ladd's masks tended to be heavier than Wood's, hers were considered superior in terms of their lifelike qualities. A recipient donning one of Ladd's masks was purportedly able to hold a crowd of surgeons guessing whether his eyes were real or painted. And many of her masks were modeled with the lips slightly open to allow the wearer to smoke a cigarette. Ladd recalled how many of her patients "sat like clods in the hospitals until, protected by their masks and again able to walk the streets and be recognized by their friends, they could go back into life and renew their struggles and accomplishments. They could smoke and twirl their moustaches; their children could say: 'Here comes papa!'"

As the war progressed, the popularity of "tin nose shops" grew to reflect the number of casualties. As with surgeons, however, there were not enough artists to meet the demand for masks. When Henry Brooks, an Englishman fighting in France, lost part of his nose from a gunshot wound, he was told that his injury was too minor to warrant immediate surgical attention. Brooks, who was an optical mechanic before the war, put his skills to use and fashioned himself a nasal prosthetic out of aluminum, which he treated with acid to give it the slightly pitted appearance of human skin. He then painted his new nose in flesh tones. The result was so successful that Brooks began producing similar prosthetics for other war casualties.

Yet, for Gillies, the mere existence of the masks was a stark reminder of the limitations of plastic surgery. Wood observed that "the cases that come to me are those in which the wounds . . . have been so severe as to remove them beyond the range of even the most advanced plastic surgical operations." Gillies found that he often had no choice but to recommend masks to some of the men in his care. One patient, Rifleman Moss, had lost both of his eyes and a large portion of his nose and upper jaw. When Gillies went as far as he could with the reconstructive surgery, he had the soldier fitted with a mask that was held in place by a pair of dark glasses.

When a mask had to be used, Gillies preferred it to be a temporary solution for convalescents awaiting further surgery. A South African patient wore a mask whenever he was granted day leave to visit friends and family in London. If the weather was hot, he would be forced to remove his metal prosthesis due to the heat. When the young soldier returned to the hospital after leave, he would hold up two, three, or sometimes four fingers to signal the number of people on the streets who had been startled by his underlying appearance. Occasionally, ladies on the bus "collapsed in a faint with horror or wept with sorrow as they looked at him."

Gillies harbored serious doubts about whether masks could offer a long-term solution to his patients. Although they were highly rendered—even beautiful—their unchanging expressions could be unsettling to onlookers. Put simply, masks could never register emotions the way a human face did. When one of Wood's patients was granted leave to visit his family in south London, his children fled in terror at the sight of his frozen features. Gillies understood this all too well. "One can appreciate a sweetheart's repugnance at being expected to kiss shapely but unresponsive lips composed of enamelled phosphor-bronze," he quipped. Furthermore, the mask could not age with its wearer, thus rendering it obsolete after a time.

In addition to aesthetic limitations, the masks were uncomfortable to wear and difficult to secure to the head. They were also fragile. Over time, the paint would flake off and the metal would begin to tarnish. Of all the masks that Anna Coleman Ladd designed during the war, none is known to survive today. Although the goal of the mask was to restore a soldier's dignity and ease his transition into civilian life, the mask itself served as a reminder of its ultimate purpose: to conceal. This was done mainly for the benefit of the viewer. Lon Chaney donned a mask not dissimilar to those made by Wood and Ladd to cover his character's distorted features in the 1925 movie adaptation of *The Phantom of the Opera*. Indeed, there was something haunting about these masks. For these reasons, many soldiers turned against them. "These blankety tin faces are no good to us," they would complain. "Can't you give us something that we can wash and shave and won't fall off in the street?"

In the end, most disfigured men were willing to subject themselves to painful experimental operations in order to restore their appearance. "We know from a considerable experience the patient . . . will undergo untold hardships to be restored to the normal," one field surgeon wrote. Gillies found the same to be true. He witnessed the immediate effect on the men's mental well-being: "once we started their repair their morale usually led the pace, as evidenced by many a moustache perking up with a bit of spit and twist." Moreover, the range of what Gillies could offer was expanding, as each new phase of the war produced greater numbers of damaged men upon whom he was required to hone his skills.

An endless flow of casualties continued to stream into Aldershot while the Battle of the Somme raged on. By the end of the summer, the Cambridge Military Hospital seemed to have reached critical mass. Gillies knew he needed more space to accommodate patients. "My days and nights were filled with the problems of the wounded," he despaired. Despite the fact that Sir William Arbuthnot Lane had allocated Gillies a further two hundred beds, there was still not enough room to house all the injured men arriving from the Western Front.

The problem was compounded by the fact that those undergoing facial reconstruction often required multiple operations interspersed with long periods of recovery. As a result, Gillies's turnover was much slower than the average surgeon's—though he sometimes gave in to external pressure from the men themselves to reach prematurely for the scalpel. "I found myself operating too soon and too often," he confessed. "It takes time to build back a face, but these mutilated young warriors, bored sitting around waiting for their next operation, urged me into surgery long before their haemoglobin and scar healing were up to it." Inevitably, mistakes were made, and some patients experienced setbacks. "It would have been better to fit them with moustaches and let them rest for a time," Gillies reflected. The value of never doing today what one could put off till tomorrow was continuously emphasized by his work at Aldershot.

With the overcrowding problem escalating rapidly, Gillies approached Lane with the idea of developing a large convalescent facility where his patients could recover from surgeries. This would free up beds at the hospital and make room for new arrivals. Lane agreed and enlisted the help of a minor aristocrat named Lady Rodney, who offered her country estate at Great Alresford in Hampshire to convalescents. While this new facility provided some relief to Gillies and his staff, more space was still needed to accommodate the deluge of casualties. The Cambridge Military Hospital was simply not equipped to cope with the mass carnage produced by a full-scale industrialized war.

Gillies badgered the "brass hats" at the War Office for a solution, but to no avail. Frustrated once again with the government's unwillingness to act, he appealed directly to the British Red Cross. With their help, a committee was formed, and his pleas were finally heard. The War Office granted permission for his unit to be moved to new and larger premises.

The newly formed committee approached Charles Kenderdine, a well-connected land agent, for guidance on suitable sites. Kenderdine alerted them to Frognal House. He had been acting as estate agent for the property ever since it was put up for sale in December 1914 following the death of its owner, Robert Marsham-Townshend. Situated twelve miles southeast of London in the town of Sidcup, the early-eighteenth-century mansion sat on 1,740 acres of land. It was close to the main rail line to Dover, thus ensuring a direct link to France for incoming casualties. It was the perfect site for the establishment of a new hospital.

Kenderdine was made honorary secretary and treasurer of a committee charged with raising funds to secure the land. Generous donations began pouring in from organizations and individuals alike. Queen Mary was among the many illustrious benefactors, and she eventually lent her name to the hospital to encourage further donations. It wasn't long before Kenderdine and the committee were able to lease Frognal House and the surrounding grounds. Construction of the Queen's Hospital began in earnest in February 1917.

As he did with the specialty unit at Aldershot, Lane played a role in helping to establish the new hospital at Sidcup. He wrote to Gillies about his aspirations: "I want to make Sidcup the *biggest* and *most important* hospital for jaws and plastic work in the world and you consequently a leader in this form of surgery."

But not everyone would follow Gillies to the new site. Nurse Catherine Black, who had worked tirelessly alongside him at the Cambridge Military Hospital, was sent to France before the hospital was up and running. There, she worked to rehabilitate officers suffering from shell shock, describing it as "one of the saddest conditions of modern warfare." Shell shock, or what modern clinicians might call post-traumatic stress disorder (PTSD), was so pervasive during the First World War that the term has almost become synonymous with the conflict itself. At the beginning of hostilities, between three and four percent of soldiers of all ranks were being evacuated from the front due to "nervous and mental shock." The psychological effects of combat were little understood at the time. Doctors initially assumed that the disorder stemmed from the concussive power of artillery barrages—hence the term "shell shock." Such fundamental misunderstanding betrayed the inadequacies of medicine in treating this condition. Nurse Black often felt just as useless to this type of traumatized soldier in France as she did to the disfigured men in Aldershot.

But at least progress was being made in the treatment of physical wounds. Under Harold Gillies, the Queen's Hospital would soon attract some of the top surgeons and dentists from around the world. They would travel there to observe the latest innovations in plastic surgery, of which there were many. Gillies would blaze a trail for a new generation of plastic surgeons—practitioners who were concerned not just with function, but also with aesthetics. His techniques would help restore the faces of thousands of injured men.

The writer Reginald Pound remarked that the Cambridge Military Hospital in Aldershot was the prenatal clinic of modern plastic surgery, and its birthplace was the Queen's Hospital in Sidcup. It was there that many of the principles of contemporary plastic surgery were established before they were eventually adopted worldwide.

Gillies arrived in Sidcup on August 18, 1917—five months after construction on the property had begun—and started operating immediately. "We literally put down our suitcases and picked up our needle-holders," he wrote. "Is there a better way to open a hospital?"

❊ 8 ❊

THE MIRACLE WORKERS

illies beckoned for the bewildered man loitering in the
doorway to come closer. "These spots here are the eyes,"
he explained to Harold Begbie, a journalist who was visiting Sidcup in an official capacity. Begbie had been reporting on
the medical community's response to the war since the conflict
began, and as a result, he had witnessed the best and the worst of
humanity over the last few years.

"War is horrible, devilish, and unutterably loathsome," Begbie
reported in the *Liverpool Daily Post*. On an earlier visit to a makeshift hospital near the front, he had met a young man who had been
shot repeatedly in the head and yet survived. Surgeons were able
to remove four of the bullets, but two remained lodged firmly in
his skull—a lifelong reminder of how close he had come to death.
On another occasion, in the more salubrious setting of a British
hotel lobby, Begbie had overheard an exchange between an eminent bacteriologist and a prominent surgeon who were enjoying
drinks together. The bacteriologist turned to his friend, gestured
at the uniformed soldiers buzzing about the room, and expressed
the brutally simple calculus of medicine in wartime: "Here we are,

you and I, whose business it is to save life, in the midst of men whose business it is to destroy life."

Now, in the operating room of the newly constructed Queen's Hospital, Begbie was a witness to that very destruction. As he inched closer, the patient came into focus. He was naked to the waist, and his body was smeared with iodine, giving his skin a lurid orange tint. Gillies used his scalpel to point to an area on the man's chest that bore the faint, hand-drawn outline of a face. "[T]his is where the nose will go," he explained to Begbie, "and here you see the mouth we shall give him." Begbie was both horrified and entranced. He could not look away. Writing about the experience later for the *Yorkshire Evening Post*, he recounted, "I can see the patient is a man, and I can see that once upon a time this man had a face: but I am thinking not of . . . the damnable wickedness of war; only how long I shall be able to stand looking at this dreadful creature who is still a man."

If the observer looked upset, Gillies took no notice. A mutilated face barely gave him pause these days. But Begbie was not inured to such sights. The journalist was especially struck by the haunting spectacle of "[t]hat pencilled face on the man's chest, like a mask and above [it] . . . the old blasted and shattered face that a few days ago had the beauty and freshness of youth." He wondered who the anesthetized man lying at the center of the room had been before the war, and who he would become after this series of painful operations was complete.

Gillies broke Begbie's trance by grabbing the patient's shoulder to adjust his position. Anticipation filled the room as he positioned his scalpel and made the first cut. At that moment, another staff member whispered to Begbie, "You see those little swellings on the shoulder? Those are bits of bones which have been taken from the man's ribs and placed there to form the cartilage of the nose." With appalled fascination, Begbie inspected the two distinct ridges formed of bone, implanted there three days earlier.

Working deftly with his blade, Gillies began raising skin from the patient's chest for transplantation onto his featureless face. As he did so, the same staff member continued to explain: "[T]he whole face on the chest will be lifted up and placed over the disfigured face; the nose will be built up with the cartilage taken from the ribs—it will be lined with the real living skin; the tissue, fed naturally by blood, will grow in its new place like a graft; and then all the scars will be removed."

Gillies was fully absorbed in his work when Begbie was suddenly overcome with a desperate urge to escape. The journalist was escorted from the room to an open window and offered a cigarette. Although what he had just witnessed was gruesome, Begbie understood its significance. "Mr. Derwent Wood, the most imaginative of our English sculptors . . . made masks for disfigured soldiers, so wonderful that across a room they looked natural. But now surgery is its own sculptor," he later reflected.

While he stood in the corridor trying to regain his composure, someone handed him an album containing photos of patients before and after their operations. Begbie was mesmerized by these surgical transformations. "A revolution has come," he marveled. "A new face is grafted on, and grows there, and becomes a real face— not a mask that hides horror." Looking at the photos of countless soldiers whose quality of life had been improved through plastic surgery, Begbie couldn't help but think of the work being done by Harold Gillies and his team as "a miracle."

The Allies were not the only ones performing such "miracles." The Germans were also making strides when it came to the management and treatment of facial injuries. From June 1916 to January 1922, the German Jewish surgeon Jacques Joseph ran a division at the Charité hospital in Berlin not dissimilar to the one Gillies had established at Aldershot.

Joseph was no stranger to disfigurement. He bore a scar on his left cheek from his saber-wielding days as a university student in the 1890s—though he may not have minded, for in Germany, dueling scars were visible proof of a man's courage. Because of this, university students joined dueling fraternities and challenged each other as a matter of course. The duelists protected their eyes and throat but left their faces exposed to the blade. Very little medical attention was given to any slashes or gashes sustained during the duels. One observer reported, "On purpose [the wound] is sewn up clumsily, with the hope that by this means the scar will last a lifetime."

Young Germans sought out the "honorable scar" well into the twentieth century, as illustrated by the case of a lottery winner in the 1920s who asked a surgeon to create an artificial dueling mark on his face. He wanted to be able to "pass" as someone worthy of being challenged to a duel. When the surgeon refused, the man went to a barber, who happily agreed to slash his cheek in exchange for a sum of money. Unfortunately, the barber also severed the man's salivary glands in the process.

The idea of the "honorable scar" may have softened the psychological impact of disfigurement for the German soldier. But the facial wounds many soldiers received were far worse than simple lacerations, and it wasn't long before Joseph's new unit at the Charité hospital was overrun with patients requiring urgent care.

Unlike Gillies, Joseph had experience in reconstructive surgery prior to the war. In the 1890s, he performed a surgery to pin back the large, protruding ears of a ten-year-old boy who refused to attend school due to the ridicule he received from his classmates. Although the surgery was a success, Joseph lost his job as a consequence. His superiors worried about the impact these sorts of experimental procedures might have on the hospital's reputation. Surgery, they argued, should not be performed for vanity's sake alone. But the incident did not deter Joseph from his growing interest in cosmetic surgery.

Afterward, Joseph joined a private practice and began performing other, similar procedures, including rhinoplasties for Jewish
clients wishing to modify the perceived facial signifier of their ethnicity. (Joseph had undergone an alteration of sorts himself, changing his name from Jakob to Jacques while he was a student.) At a
time when most surgeons scorned such "frivolous" procedures, Joseph asserted that the psychological impact of plastic surgery was
as important as its ability to restore function. It was a philosophy
that would serve him well when he took on the task of rebuilding
the faces of soldiers injured during the war.

In many ways, Joseph's work mirrored Gillies's. Both men were
interested in form as well as function. Joseph won acclaim after he
reconstructed the face of a solider named Musafer Ipar. The un-
lucky Ipar had been severely disfigured when the Allies launched
an attack on Turkish forces during the Gallipoli campaign in an
ill-fated attempt to take control of a strategically located strait sep-
arating Europe from Asia. Ipar's cheek, lips, nose, palate, and right
orbit had been completely obliterated. Given the seriousness of his
injuries, the Red Cross flew Ipar to Berlin, where he was admitted
to the Charité hospital. There, Joseph worked tirelessly to rebuild
the middle third of the soldier's face.

In many cases, Joseph's work was nothing short of miraculous.
Like Hippolyte Morestin in Paris, however, he eschewed collaboration. When surgeons visited the Charité to observe him in the
operating theater, they were not allowed to ask questions and were
often met with cold indifference. This was in stark contrast to Gillies, who was working hard to assemble a multidisciplinary team
at Sidcup that would include surgeons, physicians, dentists, radiologists, artists, sculptors, mask-makers, and photographers—all
of whom would assist in the reconstruction process from start to
finish.

•••

*Gillies knew that collaboration was key to the development of new tech-*niques in plastic surgery. His vision for the Queen's Hospital aligned with Sir William Arbuthnot Lane's, who continued to impress upon him the need to make Sidcup the preeminent site for jaw and plastic work in the world. "The larger the hospital, the bigger the men associated with it, the stronger will be your position," Lane wrote to Gillies.

Gillies had organized the hospital with the same careful attention to detail that he exhibited in the operating room. Frognal House had been repurposed to provide administrative space, accommodations for nursing staff, and a mess for convalescent officers. Also housed inside the walls of the eighteenth-century mansion was an art studio for Professor Tonks, who by then was splitting his time between the Slade School of Fine Art and Sidcup. Beyond the historical building was a series of wooden huts that acted as wards, each one containing twenty-six beds. These were arranged in a horseshoe configuration around a central admissions block. Each ward opened out onto a veranda. As one looked into the mouth of the horseshoe, septic cases were housed on the right. Moving counterclockwise, each subsequent ward accommodated progressively less-septic patients. Within the perimeter of this semicircle were multiple surgical and dental theaters—as well as examination, X-ray, physiotherapy, and photography rooms. When the journalist Harold Begbie visited, the Queen's Hospital could accommodate 320 patients. In time, the space would grow to house over 600.

Gillies and his family took up residence at Twysdens, a large Victorian house in Foots Cray, only a few miles from Sidcup. This afforded Kathleen and the children privacy, while also providing Gillies respite from the intense work occupying him at the hospital day in and day out.

From the start, it was important to Gillies that wartime surgeons specializing in maxillofacial injuries could train and work in the same place. "[N]ot until the organisation of the new home . . . has there been opportunity for anything but disjointed study in this department of surgery," he observed. At Sidcup, this would all change.

The hospital was to act as a hub for the treatment of casualties of Britain and the Dominion nations of Canada, Australia, and New Zealand, and it was divided into four sections accordingly. Each acted autonomously and was staffed by its own team of surgeons, dentists, radiographers, artists, photographers, and modelers. Gillies headed the British section and was aided in his work by the dental surgeon William Kelsey Fry—a medical officer who had been assigned to the Queen's Hospital after being wounded at the front. Early in the war, Kelsey Fry had delivered a man with a facial injury to a regimental aid post, only to discover that the soldier had suffocated shortly after medics placed him on his back on a stretcher. It was a lesson he would not forget. Kelsey Fry was put in charge of the hard tissues, while Gillies worked on the soft tissues. The British section was the largest of the four, constituting two-fifths of the patients at Sidcup. Gillies was also appointed Chief Medical Officer, responsible for overseeing the running of the entire hospital.

The Australian section, which held a fifth of the patients at the Queen's Hospital, was led by Lieutenant Colonel Henry Simpson Newland, a wiry man of a serious disposition. Born in 1873, Newland was very much a late Victorian in outlook. The Australian artist Daryl Lindsay—who had come to Sidcup to work for Newland in the same capacity as Tonks worked for Gillies—said of him: "He was a disciplinarian and kept everybody on their toes. He never spared himself and expected everybody to keep up with him." Newland was the oldest and most senior-ranking officer of the four section commanders. Nonetheless, he was content for Gillies to continue in his role as Chief Medical Officer due to his extensive experience in maxillofacial surgery, which far exceeded

Newland's own. (Unlike infantry units, military hospitals set more store by the skill of doctors than by their rank when determining leadership roles.)

Next, there was Major Carl Waldron, who led the Canadian section. Waldron had trained in both otolaryngology (ear, nose, and throat) and maxillofacial pathology, which made him uniquely qualified to handle jaw and face wounds. At the onset of war, Waldron had tried to enlist in the Canadian Army Medical Corps but was frustrated to discover that there were nearly four hundred applicants ahead of him. So eager was he to join the war effort that Waldron paid for his own passage to London, bringing with him a letter of recommendation from Sir William Osler, a prominent Canadian physician and founder of Johns Hopkins Hospital, who revolutionized methods of medical education. It wasn't long before Waldron secured a commission as a lieutenant in the British Army. From 1916 onward, he specialized in the treatment of facial injuries to Canadian casualties, first at the Westcliffe Canadian Eye and Ear Hospital in Kent and later at the Ontario Military Hospital in Orpington, just six miles from Sidcup. Waldron was assisted in his work by Fulton Risdon, who, like Kelsey Fry, was trained in both surgery and dentistry.

The last, and most reluctant, officer to be brought on board was Major Henry Percy Pickerill. He was the former dean of the University of Otago dental school and was put in charge of the New Zealand section—though Gillies could arguably have taken charge of this unit, being a New Zealander himself. Unlike the other three section chiefs, Pickerill was a dentist with no formal training in surgery. After he arrived in England in March 1917, he was sent to No. 2 New Zealand General Hospital at Walton-on-Thames in Surrey. Initially, he wrote that he was put in charge of a general ward that contained "all medical and surgical cases of every conceivable type, so that of necessity I had to furbish up my surgery and medicine." Over time, he began to specialize in jaw and face injuries, becoming a pioneer in bone grafts and transforming himself from a university dean into a maxillofacial surgeon.

Pickerill had initially resisted the relocation of his patients to Sidcup, as he believed he could best serve them where he was stationed. When Queen Mary visited No. 2 New Zealand General Hospital, however, she expressed surprise that his unit hadn't been moved. The queen—who had helped pave the way for part of the Royal Mews at Buckingham Palace to be converted into a medical ward, and who visited wounded and dying servicemen in the hospital on a regular basis—discussed Pickerill's concerns with him. But in the end, she made her wishes plain. "Well, Major, to please *me* I would like you to go to Sidcup," she told him. Unsurprisingly, Pickerill and his twenty-nine patients moved to their new premises shortly after the queen's visit.

In addition to these four men, a number of surgeons from the United States were stationed at Sidcup. They were not given their own fiefdom, but rather dispersed throughout the hospital. This was done in anticipation of a likely flood of American soldiers in need of reconstructive surgery, since the United States had entered the war several months earlier, in April 1917. The nation's decision was prompted in part by the interception of a telegram from the German Foreign Secretary Arthur Zimmermann to the German ambassador to Mexico, Heinrich von Eckhardt. In his message, Zimmermann instructed Eckhardt to propose an alliance between Germany and Mexico. As part of the arrangement, the Germans would support the Mexicans in regaining Texas, New Mexico, and Arizona—territories they had lost during the Mexican-American War in the mid-nineteenth century. News of the telegram sent shock waves through the United States. Shortly afterward, President Woodrow Wilson went before a special joint session of Congress and asked for a declaration of war against Germany. Two months later, the first American troops—dubbed the "Doughboys"—landed in Europe.

Although such developments were giving Gillies's team a very international flavor, there were some notable absences at the hospital. When it first opened, Gillies asked the dental surgeon Varaztad Kazanjian to join his team. Word had spread that this Armenian American was performing miracles on the faces of injured soldiers in France. But Kazanjian declined Gillies's invitation. "I had been sent to France by Harvard, and had responsibilities," he wrote. In the end, he felt that he could do his best work closer to the front, "with his feet in the mud." The two men nevertheless maintained a correspondence, and many of Kazanjian's patients ended up in Gillies's care once they were transferred back to Britain. "As the years of the War continued, I visited Sir Harold frequently at Queen's Hospital at Sidcup, where many of my patients were who had been forwarded for further reconstructive surgery," the dental surgeon recalled. It's unclear whether Gillies ever asked Auguste Charles Valadier to join him, but—as with Kazanjian—he kept in touch with the eccentric French dentist throughout the war and over time also received some of Valadier's patients.

Back at Sidcup, Gillies hoped that the establishment of a single-site, specialist hospital would encourage cooperation among the various practitioners working there. Gillies boasted of the diversity at Queen's: "This was indeed an impressive array . . . Out of many a heated meeting [at the hospital] floated a symphony of accents, the Canadian North Irish brogue, the New Zealand Fiji twang, the Australian cockney, a Midwestern drawl, a Philadelphia bark and a New York Oxford accent." In time, this unlikely team would advance the discipline of plastic surgery by creating techniques that could be tried, tested, and standardized. Even Pickerill, who had originally opposed a move to Sidcup, came to understand the advantages of being there. "The whole hospital was an excellent example of the harmonious interlocking of the forces of the British Empire and U.S.A.," he remarked.

In this way, what Gillies and his colleagues were able to accomplish in such a limited period was indeed nothing short of miraculous in comparison to work being done elsewhere. The Queen's Hospital even garnered the attention of the world's leading medical journal, *The Lancet*, which commented on the novelty of its approach to reconstructive surgery. "Such an intensive culture of scientific method under the stimulus of collective interest and criticism is here seen [at the Queen's Hospital] to produce results which could hardly have been attained by a generation of sporadic individual effort," the medical journal reported in 1917.

The pooling of so much talent also spurred competition among the various surgeons at the hospital—especially Gillies, a natural sportsman who never missed an opportunity to outshine a perceived opponent. Like any artist, each surgeon had his own unique style. It soon became easy to distinguish among noses reconstructed by Gillies, Newland, Waldron, or Pickerill. "With our artistic efforts constantly on exhibition about the wards, not only the patients judged our results but we, too, if only out of the corners of our eyes, jealously compared our work with that of our colleagues," Gillies confessed. As a consequence, standards rose across the board. Even Sir William Arbuthnot Lane recognized the value of these professional rivalries, observing that the "competition brought out many men who were excellent at plastic surgery and who also vied with each other in advancing this special form of surgery." The results were often remarkable.

But there was a downside. In due course, this "friendly rivalry and hearty competition," as Gillies characterized it, became a source of tension. Medical patents on surgical techniques were a rarity, since they were seen to be at odds with the moral mission of the profession, which prioritized patient care over self-interest. If financial rewards were difficult to obtain, surgeons at least sought recognition for their innovations. In such an environment, disputes broke out over who had first devised certain procedures. Pickerill often clashed with Gillies on these matters. Long after the war, he preempted one such debate by sending Gillies an article he had written. In his own files was a note that read, "Reprint extracted and sent to Gillies . . . So that there should be no doubt as to who first skin grafted a buccal sulcus [the depression between the cheek and bone of the jaw] with open free graft under pressure."

Such is the proprietorial nature of highly ambitious surgeons, in both war and peace.

>—<

On October 3, 1917, Able Seaman William Vicarage was wheeled into the operating theater at the Queen's Hospital in Sidcup. Eighteen months earlier, he had sustained severe cordite burns while aboard HMS *Malaya* during the Battle of Jutland and had been left with extensive scarring to his face. His eyelids and lower lip were turned inside out, and so he was unable to close his eyes or open his mouth. His hand had also been severely burned. Over time, scar tissue had caused the fingers to contract, leaving him with a clawlike appendage.

Vicarage was one of the worst burn victims Gillies had ever seen. "How a man can survive such an appalling burn is difficult to imagine," he recounted. The twenty-year-old's injuries were horrifying, not to mention incredibly painful. So bleak was his situation that Gillies later wrote that "it required very considerable moral courage to attempt an operation such as could in any way radically cure the condition." But Gillies had come to realize that men who survived trauma the likes of which Vicarage had experienced often evinced an "unquenchable optimism which carries them through almost anything." As it turned out, the sailor would need just such an indomitable spirit to endure the surgeries required to repair his injuries.

Shortly after Vicarage arrived at Sidcup, Gillies decided the best way to proceed was to take flaps from the sailor's chest to replace the damaged areas of his face—a complex and painful procedure that was not dissimilar to the one witnessed by journalist Harold Begbie. Although Gillies had performed this technique on numerous patients, he knew it carried with it certain risks that were amplified when dealing with large surface areas. And yet, there were few options when it came to addressing the extensive burns on Vicarage's face, since skin grafts were less reliable.

Unlike a flap, which remains attached on one side to the original site in order to maintain its own blood supply, a graft is completely detached from the body before being transferred. Grafts survive as oxygen and nutrients diffuse into them from the underlying wound bed, but long-term survival depends on a new blood supply forming quickly. As a result, grafts had higher failure rates in earlier periods.

As with medical treatment for burns, the earliest mention of skin grafts can also be found in the Ebers papyrus. But it wasn't until the latter half of the nineteenth century that significant advances were made in grafting. A Swiss surgeon named Jacques-Louis Reverdin devised a technique for removing tiny pieces of skin from a healthy area of the body and embedding them into wounds. Reverdin's grafts were harvested by pinching the skin between the index finger and thumb and then using a sharp instrument to shave away small pieces of the epidermis without drawing blood. "Pinch grafting" (as it became known) was an important step forward. Nevertheless, this type of graft was slow to heal and often led to contractures as the transplanted skin shrank.

Several years later, the French surgeon Louis Léopold Ollier developed a different approach that utilized strips of skin placed closely together over the injury site. This procedure was refined and popularized by the German surgeon Karl Thiersch. In general, the technique resulted in faster healing, less scar formation, and less contracture than pinch grafting. Still, it was not without its problems. The grafts were cut freehand with a large knife, which made it difficult to standardize the thickness, in spite of instruments created especially for the purpose. Additionally, the precise thickness of the tissue needed for a graft to take successfully was not yet understood. For these reasons, Gillies preferred to use a flap, especially when tackling a large area of missing skin.

At Sidcup, Gillies began operating on Vicarage by cutting a V-shaped flap from the sailor's chest, which he then stretched over the burned surface of his lower face. Next, Gillies raised two thinner strips of skin from Vicarage's shoulders and prepared to attach the free end of each flap to his face. As Gillies was doing this, he noticed that the edges of the skin from the two shoulder flaps tended to curl inward, like rolled paper. "If I stitched the edges of those flaps together," Gillies wondered, "might I not create a tube of living tissue which would increase the blood supply to grafts, close them to infection, and be far less liable to contract or degenerate as the older methods were?"

As he continued operating on Vicarage, the idea took hold in his mind, and his hands began to move almost independently of conscious thought. "[A]nother needle was threaded and, in an astonished silence, I began to stitch the flaps into tubes," he recalled. Not only was Vicarage about to undergo a transformation; so, too, was the entire field of plastic surgery. "Those tubes of Seaman Vicarage became historical treasures," Gillies later reflected. "They opened the door to a bigger, finer area of development than we had ever glimpsed."

Gillies called his invention the "tubed pedicle." It was a flap of skin that was stitched into a protective, infection-resistant cylinder, the free end of which was attached to the site of injury. Unlike open flaps that left the raw underside exposed, this technique dramatically reduced the chance of infection by encasing the tissue in a protective outer layer of skin. Once a blood supply had been satisfactorily established at the new site, the original connection could be cut.

"I could bring them from one part of a patient to another, in easy stages," Gillies enthused. There was little he could not do with the tubed pedicle. "I could make them in the form of a 'U' with both ends still attached to their base, allowing the blood vessels inside to become adjusted until needed," he explained. Before long, dozens of soldiers on Gillies's wards had trunk-like tubes sprouting from their foreheads, cheeks, noses, lips, and ears—all carrying with them the promise of the miraculous reconstruction of increasingly disparate parts of the body.

It wasn't long before Gillies's colleagues adopted his technique. "The enthusiastic rabble of surgeons pounced on this method," he wrote. The wards of Sidcup soon "resembled the jungles of Burma, teeming with dangling pedicles." Ten years after the war, Gillies encountered one of his patients who still had a trunk-like nasal pedicle in place. Gillies had been forced to send him home to await further operations when a new German offensive put too much pressure on the hospital. In the midst of this chaos, the man had been forgotten. When Gillies asked him about his life in the intervening years, he remarked (possibly in jest) that he had been earning a living as an "elephant man" in a traveling circus. But for all the unedifying voyeurism they attracted in the outside world, these appendages became emblems of the innovative work being done at the Queen's Hospital. "If all the tube pedicles that I have made and those my assistants have made were laid end to end, by calculation at two and a half pedicles per week, they would string like sausages from Buckingham Palace down the Mall, straight on through the Admiralty Arch to Trafalgar Square and half-way up Nelson's monument," Gillies joked. "It is my ambition that before my last pedicle is made we will reach the top of this famous pinnacle with at least one pedicle left to go into the Admiral's palate."

An early case involving the use of the tubed pedicle was that of a patient who not only shared Gillies's surname, but also held the rank of major. Inevitably, this led to confusion, and Gillies often received the man's letters by accident. One day, Major T. Gillies's laundry bill was delivered to Harold Gillies in error. Jokingly, the surgeon grabbed his namesake by the collar and growled, "I don't mind reading your love letters but I refuse to pay for your dirty linen."

Successes with tubed pedicles led to the method becoming so favored that, as Gillies later admitted, they were sometimes used to the detriment of the patient, when local or other types of skin flaps would likely have been more effective. "As in all innovations, limitations . . . had yet to be discovered, and in the process the pendulum was allowed to swing too far," Gillies confessed.

Gillies and his team were making huge strides in a previously neglected field of surgery. But the joy of successful innovation was frequently tempered by reminders of the war. One evening, Gillies picked his way through the gathering dark to Twysdens, where Kathleen and his children—John, age five, and Margaret, age three—were waiting to welcome him home. He relished the brief solitude these walks afforded him after a strenuous day's work in the hubbub of the hospital. Lately, the world had been weighing heavily on his shoulders—so much so that, for once, he barely took notice of the sinister silhouettes of the German airships overhead. They were heading for London, and their steel underbellies were stuffed with incendiary bombs and grenades.

The pilots' mission was part of a prolonged aerial bombing campaign waged against Britain by the Germans during World War I. The initial attack on the capital occurred on May 31, 1915, when a door opened underneath a futuristic Zeppelin hovering over the sleepy city. The scene was eerily similar to one H. G. Wells had described years earlier in his book *The War in the Air.* As the city rocked under the bombardment that night, unsuspecting civilians leapt from their beds and ran, panicked, into the streets. In East

London, a bomb hit the house of Samuel Leggatt, injuring four of his children and killing his three-year-old daughter, Elsie. From that day forward, the Zeppelins were known as "baby killers."

For many Londoners, the war was no longer "over there" but right here on their doorsteps. Doris Cobban wrote of being five years old and awoken in the middle of the night by the bombs. She remembered her father coming up to the bedroom. He picked her up, wrapped her in a blanket, and told her, "[T]his is history, you must see this." Indeed, British civilians were beginning to understand what their fathers, sons, brothers, and husbands fighting on the front already knew: War did not discriminate. Nobody was safe. Everyone was a target.

These monstrous hydrogen balloons with their cigar-shaped steel frames—twice the length of a modern jumbo jet—were particularly effective at surprise attacks. At eleven thousand feet, the Zeppelin could turn off its engine and drift silently toward its targets, far above the reach of most biplanes. Antiaircraft fire only forced the airships higher. And searchlights that were designed to scour the sky for potential threats proved largely useless, except in causing apprehension among Londoners. "You had about as much chance of spotting a black cat in the Albert Hall in the dark," joked one man.

Still, some pilots got lucky. Reginald Warneford became the first airman to shoot down a Zeppelin after spotting one above Ghent, Belgium, on June 7, 1915. The bullets from his biplane's machine gun tore through the flanks of the floating beast but were not enough to bring it down. Warneford chased the Zeppelin, switching off his engine and gliding above it before dropping six incendiary bombs on top of it. The force of the exploding hydrogen spun his own airplane upside down, forcing him to make an emergency landing in enemy territory. Warneford worked frantically to fix his plane and took off just as the Germans figured out what was going on. He allegedly shouted, "Give my regards to the

Kaiser!" to the enemy as his aircraft left the ground. But Warneford's victory was the exception, not the rule, during those early years of the war when the limitations of early aircraft prevented the British from downing these aerial leviathans.

The British people were outraged by the attacks. Far from destroying morale, the raids unified Londoners. After one particularly deadly raid on the city, King George V pointed from a window of Buckingham Palace at the statue of Queen Victoria and cried: "The Kaiser, God damn him, has even tried to destroy the statue of his own grandmother!"

Alongside outrage, however, there was also alarm. As the bombings escalated, people began taking cover in the London Underground. Authorities ordered blackouts throughout the city and even drained the lake in St. James's Park to prevent the moonlight from reflecting off the water and directing Zeppelins to Buckingham Palace. The government also began focusing on the development of aircraft that could not only reach higher altitudes, but were also armed with two kinds of ammunition: explosive bullets that could tear through an airship's gasbag, allowing oxygen to mix with hydrogen, and incendiary bullets that could ignite the gas mixture, causing the Zeppelin to explode.

On September 2, 1916, the Royal Flying Corps was given the opportunity to put its new technology to the test when sixteen Zeppelins massed in the night sky over London. It was to be one of the largest air raids to date. As bombs began to fall on the capital, Lieutenant William Leefe Robinson clambered into his cockpit and tore off toward the enemy. Searchlights from the ground picked out Robinson's plane as he climbed to eleven thousand feet and began raking a Zeppelin with bullets. Very soon, the airship caught fire. Thousands of Londoners who had flooded onto the streets to watch the pursuit began singing patriotic tunes as the Zeppelin fell from the sky. It crashed in a field next to the house of young Patrick Blundstone, who described the incident enthusiastically in a letter to his father. "It had broken in half . . . [and] was in flames, roaring, and crackling," the child wrote. He and his family left their home for a closer look, arriving on the scene at the same time as the fire brigade. With schoolboy relish, Blundstone described the state of the crew: "They were roasted . . . like the outside of Roast Beef. One had his legs off at the knees, and you could see the joint."

Robinson was awarded the Victoria Cross within forty-eight hours of destroying the Zeppelin—the fastest the medal had ever been presented to a soldier.

The airships had betrayed themselves as vulnerable, even fragile. Of the eighty Zeppelins used by the Germans during World War I, thirty-four were shot down and a further thirty-three were destroyed in accidents. As the war progressed, Germany began to deploy heavy biplane bombers alongside the Zeppelins carrying out air raids across Britain. By the time the war was over, the Germans had killed approximately fourteen hundred civilians and injured over three thousand more through their systematic bombing campaigns—precursors to the much deadlier attacks of the Blitz in the Second World War. (The number of German civilians who died due to the Allied blockade that restricted the maritime supply of goods to Germany was far greater, reaching between five hundred thousand and eight hundred thousand deaths by the end of the war.)

But on this particular night in late October 1917, as the Zeppelins buzzed overhead, Harold Gillies's mind was elsewhere. He was preoccupied with how best to address the eyelids of Able Seaman William Vicarage, which were still twisted inside out due to the contraction of the scar tissue of his severe burns. "The poor chap had to sleep with his eyes open, wearing a mask," Gillies remarked. The surgeon was haunted by the thought that his patient couldn't shut out the world. Vicarage's life had become a waking nightmare since the Battle of Jutland. It was hard to imagine a worse existence, although Gillies had seen many that might compare.

There was no easy solution for repairing something as delicate and flimsy as the eyelid. One of the earliest mentions of eyelid surgery dates to the first century C.E., when the Roman writer Aulus Cornelius Celsus described a procedure in which an incision was made to relax tightened skin around the eyes. In 1818, the German surgeon Karl Ferdinand von Gräfe coined the term "blepharoplasty"—from the Greek words *blepharon* (eyelid) and *plastikos* (to mold)—to refer to a surgical procedure for repairing certain eyelid deformities. When it came to addressing damage as severe as Vicarage's, however, options were limited. But Gillies was working on an idea.

He knew of a technique called an epithelial inlay, invented by Johannes F. Esser, a Dutch surgeon who had studied briefly in prewar Paris under the cantankerous Hippolyte Morestin. In 1916, Esser took a skin graft from a patient's thigh and wrapped it, raw side out, around a firm supporting material, or stent. He then inserted this package into a surgically prepared pocket at the site where the graft was needed. Once the two raw sides of the pocket and graft had adhered to each other, Esser removed the stent and opened up the graft to reveal healthy skin inside. This ensured that the wound was completely lined with epithelium—a vital tissue lining the internal and external surfaces of organs and glands.

Prior to the invention of this method, wounds that were not lined with epithelium often became infected and shrank as a result of scar formation. Esser's epithelial inlay counteracted these problems by holding the graft firmly against the wound and preventing movement and bleeding while it became established. He published his findings first in German and later in English. After reading about this method in a medical journal, Gillies decided to adapt it to his own purposes.

Several weeks after Vicarage's first operation, Gillies wheeled him back into an operating room at the Queen's Hospital and prepared to duplicate Esser's inlay, with one crucial difference. He wrapped the stent with the harvested skin as Esser had done with the raw side facing out. But instead of inserting the package *into* the wound, he attached it to the *outside*, just above Vicarage's eye, before suturing it with horsehair.

There was "doubting interest" among Gillies's colleagues as to whether his imaginative solution would actually work. Eight days later, Gillies removed the stent and the stitches, which allowed the new skin to unroll. Much to his delight, "there was shining before us the most perfect lid we had so far accomplished." It had been eighteen months since the sailor had been severely burned at the Battle of Jutland, eighteen months since he had been able to close his eyes to the outside world. But now he could finally enjoy a peaceful night's rest. "It caused some stir at Sidcup," Gillies proudly noted.

Indeed, the epithelial outlay—as Gillies named it—would be of even greater use to plastic surgeons in coming decades, as his achievements during this First World War would determine what could be done for a whole new generation of burned soldiers during the Second World War.

THE BOYS ON BLUE BENCHES

oris Maud was only eleven years old when her father first took her to visit the Queen's Hospital, not far from their family home. Every Sunday morning, they boarded a red double-decker omnibus bound for Sidcup, carrying with them dozens of packets of cigarettes to hand out to the soldiers recuperating there.

As the bus rattled along at twelve miles an hour—the speed limit in 1917—Doris peered through the condensation clouding the window. Wounded servicemen out for a walk dotted the roadside leading to the hospital, their white dressings a sharp contrast to the navy-blue uniforms all the convalescents at Sidcup wore. The faces of many were covered nearly entirely in bandages to hide the human cost of war from those who might encounter these lonely wanderers in the course of their day.

For civilians, the sight of a disfigured soldier was unsettling evidence of the mass slaughter taking place on the front. Newspapers sometimes published photos of those who had lost limbs—smiling and in the care of their nurses and doctors—but rarely featured men who had lost portions of their faces. Despite this, reporters did not necessarily sugarcoat the realities of facial injuries. One journalist for the *Daily Sketch* began his article by asking readers, "What kind of vision does your mind conjure up when you hear or see the word 'wounded'?" He continued:

> Probably, if you are an average stay-at-home civilian, a limping man in a blue hospital suit or, at worst, an indefinite huddled figure on a stretcher. But there are other wounded that the mind instinctively avoids contemplating. There are men who come from battle still walking firmly, still with capable hands, unscarred bodies, but who are the most tragic of all war's victims, whose endurance is to be tried in the hardest days, who are now half strangers among their own people, and reluctant even to tread the long-wished-for paths of home. In medical language they are classed as "Facial and Jaw Cases." Think that phrase over a minute and realise what it may mean.

Most wartime journalists characterized facial injuries as the "rudest blow that war can deal," since it deprived a man of his outward identity.

Belying such reports, many soldiers spoke stoically, even cheerfully, about their injuries in letters home. Reginald Evans—the man who expressed astonishment that Gillies treated ordinary soldiers with as much care as he did officers—wrote to his mother shortly after being wounded during a nighttime reconnaissance mission on the Western Front. Although he was unable to eat due to the severity of his injuries, Evans informed her that she "needn't have the slightest worry." Despite his assurances, however, it was clear that Evans was preoccupied with his appearance.

While on a visit to Waverley Abbey, he snickered when a minister preached in a sermon that it was everybody's duty "to cultivate these good looks and make themselves as beautiful as they could." Aware of the reaction his disfigurement might elicit back home, Evans warned his mother in another letter that she "will have to prepare to receive rather an uglier duckling than before." The lighthearted tone notwithstanding, one can't help but wonder how much of this was an act for his mother's benefit. Evans ended his letter to her by teasing, "You wait till I come swanking home with my false teeth and artificial jaw. I'll show some of you up."

Despite the public's reluctance to confront this taboo of the war, the patients at the Queen's Hospital were not confined to its grounds. Gillies and his staff encouraged those who were strong enough to take the air and go for walks in the surrounding area. They strolled through the town and crowded into the "Rest Room" on High Street. This was an empty shop that had been repurposed to serve soft drinks to patients from the hospital, since the men were not allowed to drink alcohol while convalescing. The Rest Room also shielded them from the curious stares of townspeople.

Medical personnel were acutely aware of the effect that unwelcome attention could have on their patients. Ward Muir, a corporal in the Royal Army Medical Corps, wrote about the difficulties one faced when interacting with a disfigured soldier. "He is aware of just what he looks like: therefore you feel intensely that he is aware that you are aware, and that some unguarded glance of yours may cause him hurt," he observed. Muir was cognizant of how his own discomfort might impact others. "This, then, is the patient at whom you are afraid to gaze unflinchingly: not afraid of yourself, but for *him*," he added. But most people were not as self-aware as Muir and found it difficult to moderate or conceal their reactions. Moreover, the limited exposure the public had to disfigured faces, whether in newspapers or in person, only served to heighten anxieties about encountering them.

Horace Sewell, a brigadier-general who spent four and a half years at the Queen's Hospital undergoing twenty excruciating operations, remembered being sent to a nearby convalescent home in

Burnham-on-Crouch to recover between procedures. "The good people of that place requested the home to keep us in as we gave them the shivers," he recalled. Even the Prince of Wales could not disguise his unease in the presence of those with facial wounds. On a visit to Sidcup, the future King Edward VIII was refused entry to the wards reserved for the hospital's worst cases. The staff worried that the experience would be too upsetting for someone who had not previously come into contact with such severe disfigurement. Nonetheless, the prince would not be denied his inspection. Sewell wrote that "he had his way and went in, and as far as I hear had to be carried out."

In the area surrounding the Queen's Hospital, a brutally simple expedient was devised to "protect" the public. Certain outdoor benches were painted bright blue and reserved for the sole use of patients from Sidcup. When passersby saw that one of these seats was occupied, they would know to avert their eyes. The men's navy-blue hospital uniforms also served as warnings to the public. Unfortunately, this further contributed to the othering of disfigured people and must have made some patients feel even more isolated during their time there.

Laden with gifts of cigarettes, bus passengers Doris Maud and her father were among the few civilians who never shunned the disfigured soldiers, driven as they were by a sense of patriotic duty. The smallest acts of civilian kindness often made the greatest impact on the morale of those coming to terms with their injuries, and no one was more aware of this than the chief surgeon of the Queen's Hospital.

Harold Gillies stood in the alleyway between wards two and three, clutching a golf club. He often came there to clear his mind after spending hours in the operating room. The stress of having to rebuild faces day in and day out without so much as a single textbook

Harold Gillies in uniform, 1915 (*Courtesy of the British Association of Plastic, Reconstructive and Aesthetic Surgeons*)

Harold Gillies at the French Open Golf Championship, Chantilly, October 1913 (*gallica.bnf.fr / Bibliothèque nationale de France*)

The Cambridge Military Hospital, Aldershot, immediately prior to the outbreak of the First World War (*Trustees of the Museum of Military Medicine, Aldershot*)

Photogravure of Hippolyte Morestin (*Wellcome Collection, Public Domain Mark*)

Illustrated map of the Queen's Hospital, Sidcup (*Courtesy of the British Association of Plastic, Reconstructive and Aesthetic Surgeons*)

Bandaged patients with nurses outside the Queen's Hospital, Sidcup (*Courtesy of the British Association of Plastic, Reconstructive and Aesthetic Surgeons*)

THE QUEEN'S HOSPITAL, "FROGNAL", SIDCUP. XMAS, 1917. WARD

Staff and patients celebrate Christmas at the Queen's Hospital, Sidcup, in 1917. (*Courtesy of the British Association of Plastic, Reconstructive and Aesthetic Surgeons*)

Henry Tonks in his room at the Queen's Hospital, Sidcup (*Courtesy of the British Association of Plastic, Reconstructive and Aesthetic Surgeons*)

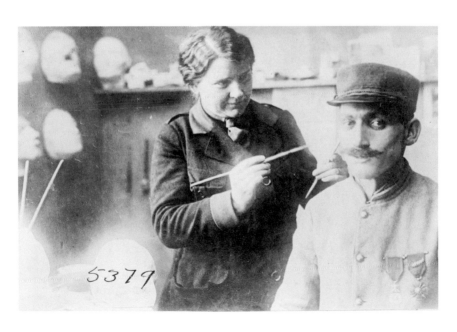

Anna Coleman Ladd in her studio, painting a mask worn by a French soldier (*American Red Cross Collection, Library of Congress; Wikimedia Commons, public domain*)

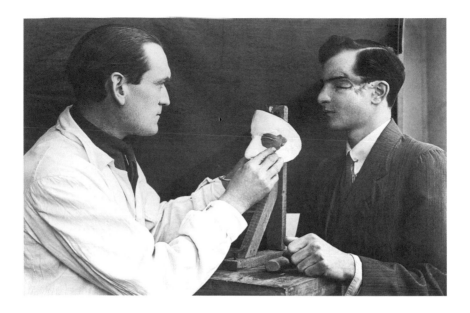

Francis Derwent Wood putting the finishing touches to a cosmetic plate and comparing it to the face of the disfigured patient for whom the plate is being made, at the 3rd London General Hospital (© *Imperial War Museums [Q30456]*)

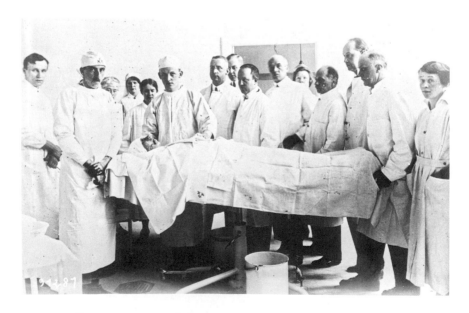

Harold Gillies, second from left, in an operating theater in 1924. He operated on Danish sailors wounded in the explosion of the cruiser *Geysir* in 1923. (*gallica.bnf.fr / Bibliothèque nationale de France*)

Percy Clare, right, later in life with his wife, Beatrice, and son, Ernest (*Rachel Gray / Photo restored by Jordan J. Lloyd*)

Percy Clare, later in life (*Rachel Gray / Photo restored by Jordan J. Lloyd*)

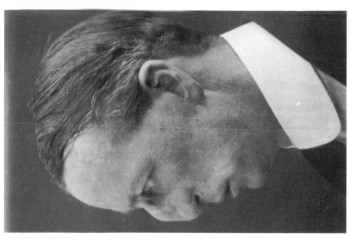

Private R. W. D. Seymour, aka "Big Bob," whose nose was partially severed on the first day of the Battle of the Somme. He eventually became Gillies's private secretary. (*From the Archives of the Royal College of Surgeons of England*)

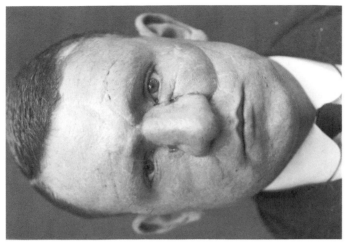

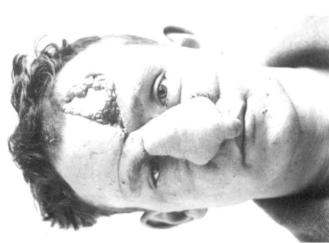

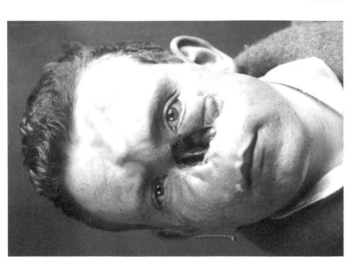

Lieutenant William Spreckley, who was admitted to the Queen's Hospital, Sidcup, on January 30, 1917. Gillies was alarmed after an early procedure left Spreckley with a nose like an "anteater's snout." *(From the Archives of the Royal College of Surgeons of England)*

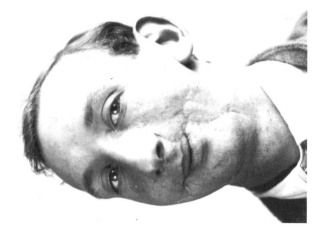
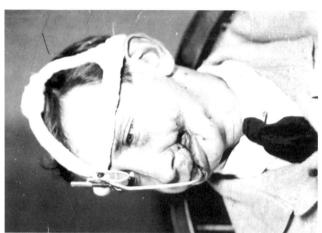

Private Walter Ashworth, who was wounded at the Battle of the Somme. Ashworth's fiancée broke off their engagement as a result of his injuries, and he later married her friend Louise Grime. (*From the Archives of the Royal College of Surgeons of England*)

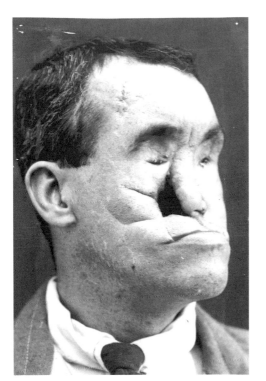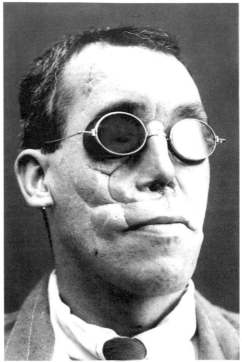

Rifleman Moss, who lost both of his eyes and a large portion of his nose and upper jaw. Gillies had him fitted with a mask that was held in place by a pair of dark glasses. (*From the Archives of the Royal College of Surgeons of England*)

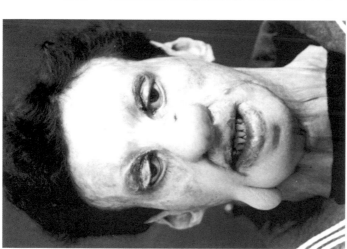

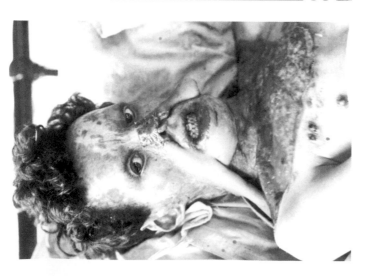

Able Seaman William Vicarage, who suffered severe cordite burns during the Battle of Jutland. Vicarage was the first patient to receive a tubed pedicle. (*From the Archives of the Royal College of Surgeons of England*)

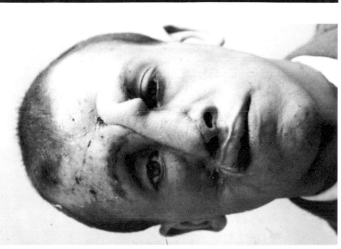

Sergeant Sidney Beldam, who was wounded during the Battle of Passchendaele. He remained where he fell for three days before being rescued. Gillies had to reopen a hastily stitched wound in order to rebuild Beldam's face—a process that took several years.
(From the Archives of the Royal College of Surgeons of England)

RIGHT Lieutenant Henry Ralph Lumley, who was severely burned after crashing his aircraft during his first solo flight. He died in Gillies's care on March 11, 1918. (*From the Archives of the Royal College of Surgeons of England*) BELOW Gillies's notes and schematic drawing detailing Lieutenant Lumley's case (*From the Archives of the Royal College of Surgeons of England*)

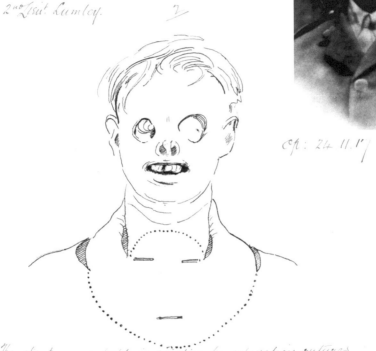

2nd Lieut. Lumley.

Op: 24.11.17

The stent was held in position by relaxation sutures
Progress: Satisfactory. On 8th day the dressings being
foul the stent was removed when it was found
that the whole of this area had been successfully
grafted in a most remarkable manner
. Melted paraffin wax was then applied over
this grafted area which unfortunately was
applied too hot. On removal of wax at
beginning the whole of the grafts were
black and subsequently none of them took.
The stent should have been left in position
two more days & then the grafted area
and exposed to the air.

15.2.18 A modification of the method of transposing
the chest flap to the face was decided on
& one extra pedicle on each side was
added to the scheme. Note photograph.

" " " General condition fair but there has been
very little attempt at regeneration of the
... area from which the neck pedicles were

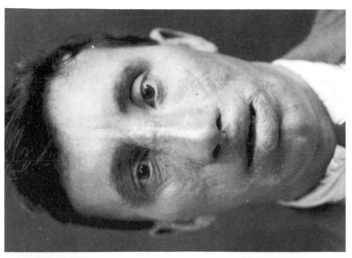

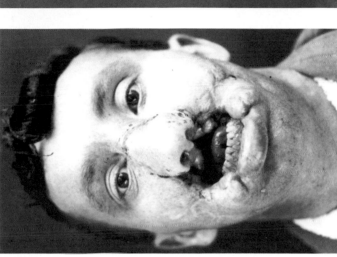

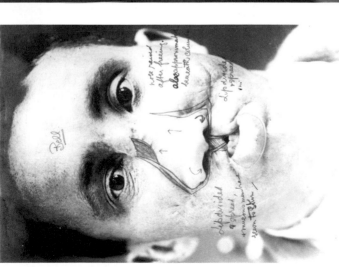

Private James Bell, who was sent to the Queen's Hospital, Sidcup, in May 1918 on the recommendation of Auguste Charles Valadier (*From the Archives of the Royal College of Surgeons of England*)

Men of the Délégation des Mutilés to Versailles, June 28, 1919. Albert Jugon is second from the right. (*FRAD048-015 Guerre mondiale, guerre totale / Europeana 1914–1918, Europe / CC BY-SA. https://www.europeana.eu/en/item/2020601/ https___1914_1918_europeana_eu_contributions_11271*)

to guide him was often overwhelming, though he was careful not to let it show. Gillies planted his feet shoulder-width apart as he swung the club in an expert arc. The little white ball whooshed down the alleyway before clack-clacking along the pavement. After a few more shots, Gillies packed up his clubs and headed back inside.

The patients at the Queen's Hospital had grown accustomed to the peculiar sight of their surgeon clanking around the corridors with a bag of golf clubs. Not only did he enjoy practicing his swing during quiet moments at work, but he also frequented the local golf course. There he often played with friends, some of whom fell victim to his practical jokes. On one occasion, Gillies managed to swap a player's ball for a replica fashioned from plaster of Paris, the same material that artists at the hospital used to make molds of the patients' faces. Upon connecting his swing with his ball, the unsuspecting player was enveloped in a cloud of white powder—much to everyone's amusement.

The sport was a welcome release for Gillies, but it couldn't always distract him from his work. After the surgeon arrived uncharacteristically late to a game one day, a friend asked him if he was feeling well. To this question, Gillies broke down and wept. One of his patients had died earlier that morning. "We played our game and I think it was the only occasion on which I scored a victory over him," his friend later recalled.

Such displays of deep emotion were a rarity for Gillies, however. It was his puckish personality and love of pranks that won the hearts of soldiers recovering at the hospital. One former patient remembered fondly that "Major Gillies himself was 'one of the boys.' He spoke their language and entered into their spirit." Gillies worked overtime to keep their spirits up during the months and years that many of them spent there. Repairing a soldier's face was difficult, but addressing the psychological damage caused by his wounds was even harder. "The injury to the subconscious mind caused by physical disfigurements is not always easily cured," Gillies wrote. But this didn't stop him from trying.

The best predictor of a patient's mental health was the outcome of the reconstructive work itself. "We noticed that if we made a poor repair for a wretched fellow the man's character was inclined to change for the worse," Gillies observed. By contrast, if the operation was a success, the patient "regained his old character and habits" and became a "happy convalescent." To Gillies, this demonstrated the impact that the physical appearance could have on a person's psyche. "[I]f my bald head suddenly flourished with a crop of curly red locks and my receding chin became thick and square, imagine how pleasant my personality would become," he joked.

Keeping his patients happy was one of Gillies's principal concerns. During the daytime, he adhered to and enforced strict hospital policies. As the sun set, however, the mischievous surgeon—often under the guise of an alter ego, whom he called "Dr. Scroggie"—would encourage the boys to break the rules. For instance, the portable gas rings that the patients sometimes used to cook meals were locked away at eight o'clock each evening. But when hunger struck (often around eleven o'clock), some of the more mobile soldiers would steal the key and cook up a feast for the ward. The smell of eggs frying often drew Gillies to the kitchen, where he would stand behind the door and jokingly shout, "Two more eggs, two rounds of toast, or I want every b[loody] name and number here." The clandestine cooks would yell back, "Two eggs here, plenty of bread, come and cook your own b[loody] supper!"

Food was bountiful at Sidcup, and the men often turned to it to alleviate the tedium of waiting for the next procedure. Captain J.G.H. Budd, who lost his nose after a shell exploded in front of him, recalled the hospital meals with fondness. "I can still look back on the breakfasts we had, and wonder how we got through them!" he wrote. "There were always two huge dishes placed on the side board, one piled high with fried eggs and the other with bacon."

Gillies believed that his patients should not be servants to routine, and as "Dr. Scroggie," he encouraged the smuggling of alcohol onto the wards from time to time. He even turned a blind eye to gambling. Philip Thorpe—a soldier who had been treated by the French dentist Auguste Charles Valadier before being sent to Sidcup—remembered teaching Gillies how to play the card game rummy. After the men had cleaned Gillies out, he took their names and jokingly threatened to report them. "However," Thorpe wrote, "we were able to prove that he had not won a game, and none of us had lost anything so it could not possibly be gambling."

Gillies's personality was undoubtedly a driving force behind the growing reputation and success of the Queen's Hospital. Donations continued to flow in from all over the country, due in part to the positive coverage the hospital received in the national press. One journalist, writing under the provocative title "SHATTERED MEN REMADE," told his readers that "[t]hese wizards of surgery will literally rebuild men's faces and transform ugliness into good looks." The newspapers urged people to dig deep into their pockets: "Will you withhold your subscription—anything from one shilling to a cool hundred or so—and let the ravages of loneliness go on among these heroes who have given all for you?" For fifty pounds, a donor could maintain the cost of a bed for a "grievously disfigured . . . hero of war" for an entire year.

The hospital continued to grow, adding a chapel, a canteen, and even a cinema. Although the buildings that made up Queen's Hospital occupied nearly ninety acres of land, Gillies still struggled to find space for the endless stream of casualties arriving from the front. A concrete structure referred to by patients as "The Jungle" was eventually built, which nearly doubled the number of beds. Added to that were convalescent beds at satellite hospitals in the immediate vicinity, which Gillies helped requisition. "No plastic unit is good unless it has an equal number of convalescent beds," Gillies argued. These auxiliary hospitals allowed him and the other surgeons to move patients on and off the main premises as needed. "By merely picking up a phone we could send one patient off for convalescence and call back another for further surgery," Gillies later wrote.

It was the autumn of 1917, and there was still no end to the war in sight. Territory was lost, won, and lost again in a deadly tug-of-war. The world's most powerful armies burrowed deeper and deeper into the earth. Both sides became more efficient at maiming and killing one another. Artillery shells were hollowed out and filled with hundreds of small steel or lead balls. Designed to detonate over the trenches, these shells wrought havoc on the bodies and faces of soldiers whose protective gear was still found wanting.

One victim of this type of shelling was twenty-two-year-old Sidney Beldam, who was seriously injured during the Third Battle of Ypres—better known as the Battle of Passchendaele, after a nearby village that bore witness to the final stages of fighting. After three months, one week, and three days of brutal trench warfare, the Allies finally recaptured the village, but at a terrible cost to human life.

In time, the name "Passchendaele" would come to evoke horrific memories in those who had been there. The Canadian doctor Frederick W. Noyes, who had also been at the Battle of the Somme, described Passchendaele as "the Somme multiplied and intensified ten times over." It was not only one of the bloodiest battles of the war, it was also one of the muddiest. Rain hammered down on the armies for nearly three months while the fighting raged. The battlefield, pockmarked by some four million bombs and shells that had been deployed during the preliminary barrage, quickly became flooded. One soldier wrote of Passchendaele that "[t]he whole earth is ploughed by the exploding shells and the holes are filled with water, and if you do not get killed by the shells you may drown in the craters."

Horses, mules, guns, and other equipment sank into deep pockets of mud. Men became trapped where they stood, unable to move or escape—making them easy targets for machine gunners. Others simply drowned. Edwin Campion Vaughan remembered hearing "the groans and wails of wounded men" who had crawled into shell holes seeking refuge, only to realize that these chasms were slowly filling with rainwater. "[T]he water was rising about them and, powerless to move, they were slowly drowning," he recalled with horror. The next morning, he saw water pouring out over the craters' edges, which accounted for the silencing of the men's cries.

Beldam found himself in the midst of this hellish landscape in November 1917. Like so many others, he soon became a casualty. A piece of shrapnel hurtled toward him, slicing through the right side of his face and tearing off a large portion of his nose. Beldam toppled face-first into the mud, and though he may not have felt lucky in that moment, he likely would have choked on his own blood had he fallen onto his back. There he remained for three days, as rats and other vermin scurried over him to nest in the corpses of his comrades—a common sight on the waterlogged battlefield. One soldier explained that the "rats were getting out of the rain, of course, because the cloth over the rib cage made quite a nice nest and when you touched a body the rats just poured out of the front . . . to think that a human being provided a nest for a rat was a pretty dreadful feeling." Because of his experiences at Passchendaele, Beldam developed a lifelong fear of rats and cockroaches.

When a group of men who had come to remove the dead approached Beldam, one soldier prodded him with a boot in an attempt to turn his body over. It was only then that they realized he was still breathing: "My god, this one's still alive." The recovery crew couldn't be blamed for thinking the crumpled, bloody heap before them was a corpse. The battlefields were littered with the dead in various stages of decomposition. "Often have I picked up the remains of a fine brave man on a shovel," one man remembered, "[j]ust a little heap of bones and maggots to be carried to the common burial place." These recovery crews faced unimaginable horrors in the aftermath of battle. "I shuddered as my hands, covered in soft flesh and slime, moved about in search of the disc [identification tag] . . . I have had to pull bodies to pieces in order that they should not be buried unknown. It was very painful to have to bury the unknown," he added.

The men gathered up the broken, mutilated soldier and carried him back to the trenches. Beldam then made the arduous journey to Britain, where he was initially sent to an auxiliary hospital at Rawtenstall, Lancashire. Due to the severity of his injuries and his overall condition, doctors gave him six months to live. Perhaps because of this, they hastily sutured up his wounds without addressing the extensive tissue loss to his face. As a result, the right side of his upper lip was permanently lifted in a snarl, while his nose appeared twisted and sunken.

But luck was on Beldam's side. Several months after he was wounded, the authorities decided to relocate him to the Queen's Hospital, where medical miracles were rumored to occur on a daily basis. On his arrival, Beldam's condition once more illustrated to Gillies why a facial wound should not be hastily closed before injuries to the underlying structure had been addressed. Although it would be painful, Gillies informed the young man that he would have to reopen the wound in order to repair the disfigurement that had resulted from the original operation. He would need to slice through the dense scar tissue, then suture a flap of healthy tissue in place to fill out the cheek, which had been largely destroyed.

For Beldam, it would be the first of nearly *forty* operations under Gillies.

Like "Big Bob" Seymour, who became Gillies's private secretary, Beldam would also find a place in the surgeon's growing entourage. He would become Gillies's personal chauffeur after the war—a job that came with its share of surprises. When Gillies forgot to renew his own driving license, he sent Beldam to the county hall to rectify the matter on his behalf, with strict orders to return promptly so that Beldam could drive him to the hospital to perform an operation later that day. When Beldam arrived at the county hall, he was dismayed to find a long line of people ahead of him. In desperation, he dropped out of the queue and begged the administrator to renew the license so he could hurry back and drive his boss to the hospital. The administrator was not amused and reproached the chauffeur, remarking that his employer shouldn't have waited until the very last day to renew his license.

In desperation, Beldam sought out Allen Daley, a known acquaintance of Gillies who happened to be working at the county hall that day. When Daley rang the administrator, he explained that the chauffeur was acting on behalf of Harold Gillies, "the renowned plastic surgeon," who was needed urgently back at the hospital. But Daley's pleas also seemed to fall on deaf ears. He was about to give up when the administrator suddenly asked, "Is he the golfer; the man who nearly won the amateur championship?" Daley confirmed that it was indeed the same Harold Gillies. "Why didn't you say so before?" the man cried. "Of course, I will do anything to oblige such a distinguished golfer. Send his chauffeur to my private office and I will put it through myself." Beldam left the county hall with a new license in his hand and plenty of time to spare.

But all of this was yet to come. For the time being, Beldam was still plodding along the road to recovery. If he felt demoralized, the arrival of a certain young woman at the hospital lifted his spirits. Her name was Winifred, and she was an accomplished pianist living in Sidcup. She had heard about the disfigured soldiers recuperating under Harold Gillies's care and decided to volunteer her musical talents to entertain the men while they convalesced. As she walked through the bustling grounds of Frognal House with sheets of music tucked under her arm, she came face-to-face with the injured Beldam. It was love at first sight.

>•<

While Gillies's patients benefited greatly from his work, not everyone saw that work as entirely original. Shortly after the invention of the tubed pedicle, a debate about who had first developed it arose, which escalated into a full-blown dispute between Gillies and his colleague Captain John Law Aymard in the years following the war.

Aymard was an English surgeon who had trained in South Africa and who had worked with Gillies at both Aldershot and Sidcup. After the war, Aymard challenged Gillies's claim that he was the inventor of the technique. In a letter to the editor of *The Lancet*, he wrote, "I would draw Major Gillies' attention to the history of the double pedicle flaps to which he attaches so much importance. The first double pedicle flap at Sidcup was performed . . . by myself." Aymard went on to detail an operation he performed in October 1917, the same month Gillies operated on Able Seaman Vicarage. This was a rhinoplasty case that he had discussed in another article, published by *The Lancet* a few months after he had performed the surgery. Aymard ended the letter with the unconvincing assertion that his motivation for writing to the medical journal was not born out of professional jealousy. "I do not intend to enter into any disputes, but hope to depict the influence of war surgery on civil practice, leaving out all war cases and much of the pettiness connected with them," he wrote.

Gillies did not let Aymard's claim go unchallenged, citing a case of his own in a letter to *The Lancet* a week later. In this response, Gillies wrote that the "operating books, surgical records, and ward sisters' books of the Queen's Hospital show the following statement of fact"—namely, that the operation to which Aymard referred happened on October 18, two weeks after Gillies had operated on Vicarage using the tubed pedicles. Thus, it was he, *not* Aymard, who should be credited with the innovation. Gillies tried to defuse the tension by ending his letter: "I am to blame in not informing Captain Aymard at the time he published his rhinoplasty case in [*The Lancet*] that he was not the first to get on to the principle of 'tubing' the pedicle." Unfortunately, in Aymard's mind, this did not settle the argument. By then, he had left the Queen's Hospital to return to South Africa, but despite the distance between them, it would not be the last that Gillies would hear about the matter from his aggrieved colleague.

Not long after Aymard wrote his letter to the *The Lancet*, Gillies learned of another challenge to his title as the tubed pedicle's originator. On a visit to America, Gillies was informed that a Russian surgeon named Vladimir Filatov had developed the technique in 1916—a year before Gillies had operated on William Vicarage. In a published article, Filatov described how he first began by experimenting on rabbits. In doing so, he learned that the circulation in a tubed pedicle is improved by the formation of new blood vessels. On September 9, 1916, he graduated from rabbits to create the first tubed pedicle on a human patient. The article contained details of this operation, along with drawings and photographs.

This news hit Gillies hard. He wrote, "I must admit that it was, at the time, a bitter blow." Nonetheless, Gillies accepted that Filatov had independently invented an identical technique prior to his own development of the tubed pedicle at Sidcup. While he was frustrated by this turn of events, he was not entirely surprised after he had given it some thought. "On the whole, the tubed pedicle was a manoeuvre that was bound to occur to any opportunist brain working in plastic surgery," he wrote.

Indeed, it would turn out that a German dentist and autodidact war surgeon named Hugo Ganzer had also developed a tubed pedicle in 1917, also independently, and without any knowledge of the work being done by either Filatov or Gillies. Given the sheer number of soldiers disfigured during the First World War, it is not surprising that three reconstructive surgeons developed the same solution to the problem of how to transplant tissue safely from one area of the body to another. In this sense, the tubed pedicle was an evolutionary rather than revolutionary development in the history of plastic surgery, arising from a great need generated by the war. It was not the first time that a medical innovation had been developed simultaneously by different people with no prior knowledge of each other's work.

Nonetheless, Gillies's colleague Aymard nursed his resentment over the next two decades. He continued to reassert his claim that he had been the first to use the technique at the Queen's Hospital in 1917. Time and again, Gillies was forced to defend himself. In a private letter to Sir Squire Sprigge, the editor of *The Lancet*, Gillies expressed disquiet over the accusations. "It is horrible to feel that Aymard thinks that I have stolen his originality in any way," he admitted.

In fact, it had long been rumored that Aymard had attended the second operation in Vicarage's course of surgeries on October 17, which was performed on behalf of Gillies by Lieutenant H. C. Malleson, and that it was during this procedure that Aymard first learned of the famed technique. Gillies confirmed he had his own suspicions when he wrote to Sprigge: "I could not publish [in *The Lancet*] what I believe to be the truth, about Aymard first seeing my case and then doing his and publishing it straight away in order, as it seemed to me, to get priority."

Aymard remained adamant his entire life that it was Gillies who had stolen the concept of the tubed pedicle. In a final letter to his unhappy colleague, Gillies wrote, "I am very sorry that this is still rankling so deeply in you. The facts, I think, are very clear . . ."

The tubed pedicle, like many reconstructive techniques to emerge from the war, evolved into a mainstay of plastic surgery under Gillies's direction at the Queen's Hospital. In later life, he wrote, "I, and I think rightly, have been credited with an independent observation of the value of the tube pedicle and all its enormous developments since." The tension between Gillies and Aymard showed that the competition between surgeons at Sidcup during the war was not always as friendly as Gillies liked to portray it. But challenges far greater than professional squabbles would test him in the coming months.

⊰ 10 ⊱

PERCY

rivate Percy Clare of the 7th Battalion, East Surrey Regiment, remained crumpled on the ground for hours while the Battle of Cambrai thundered around him. He had been shot in the face seven hundred yards away from the trench shortly after the fighting had commenced. Blood flowed from the gaping hole in his cheek, soaking the front of his uniform.

Clare had unblocked his own airway by extracting the packet of field dressings shoved there by a panicked officer named Rawson. But blood continued to flood the back of his throat, causing him to vomit every few minutes. His mind drifted to thoughts of the mass grave that awaited him once the battle was over.

Clare was largely resigned to his fate by the time the booted feet of his friend—and savior—Weyman appeared in his field of vision. Clare drifted in and out of consciousness as the men summoned by Weyman bundled him onto a stretcher and undertook the hazardous task of transporting him off the field. In his hand was a small Bible that his mother had given to him—now stained with his own blood. "[O]ur journey was most perilous still, by reason of the heavy hostile shelling," he later scribbled in his diary.

At one point, the rescue team lost their way and had to back-track through a shower of bullets. "I remember them climbing over the bank to get round the barbed wire barricade on the road which we had passed on the way up," Clare wrote. Lying there were the bodies and limbs of soldiers who had been hit by shells. As the stretcher-bearers hurriedly stepped around the corpses, one of them was shot. He jerked upright in pain, nearly spilling Clare onto the ground. But after his rescuers regrouped and picked their way through the chaos, Clare was placed onto a wheeled stretcher and escorted to an ambulance by two German prisoners pressed into duty. He recorded with gratitude that they "took every pre-caution to prevent the stretcher jolting on the uneven road." As he was loaded into the vehicle alongside seven other severely injured men, Clare waved goodbye to Wcyman. Only later would he learn that his friend had died of his own wounds shortly after the two parted ways. Clare hadn't even realized the stoic Weyman had also been injured.

The ambulance sped off, jostling the casualties inside. Every bump in the road was agony. "The driver rushed us off at a seem-ingly reckless pace, bringing awful groans and shrieks from the wounded man below me," Clare remembered. The driver could hardly be blamed for his haste. He had no time to spare, since he was trying to outrun the shells that continued to explode around them. "One dropped and burst about ten yards behind us," Clare wrote. It missed the ambulance by only a split second.

The vehicle churned its way through the mud before slowing to a halt outside a casualty clearing station. Clare was quickly off-loaded and his stretcher placed on the ground. Cannons continued to boom in the distance. He was now just one of scores of injured men lying in long rows with their boots poking out from under-neath blood-splattered blankets. Stretcher-bearers wove through this maze of writhing bodies, lifting a select few off the ground and moving them to the frenetic interior of an operating room.

One nurse remarked that "white-gowned surgeons stand so thick around the tables that you cannot see what is on them." Outside, ambulances kept arriving to off-load more and more injured men.

While Clare waited his turn, a medical officer bent over him to inspect his face. "Yes, Sir: through and through," he murmured to the sergeant nearby before moving on to the next soldier—leaving Clare at a loss to know what he meant. He had only a vague sense of his injuries. "I certainly couldn't have told them where I was wounded: I didn't know, then." Only later would Clare learn that a bullet had entered just in front of his right ear and traveled in a downward trajectory, narrowly missing his right eye and fracturing his jaw in several places before smashing through his left cheek. The sergeant scribbled out a label and pinned it to Clare's tunic. Over the coming weeks, it would be scrutinized by everyone who came in contact with him. "They merely read the label and ordered my removal from one place to another much as if I were a carcass in a meat market, already weighed, described and priced!"

Despite the severity of Clare's injuries, he was one of the lucky ones. Many casualties were captured by the enemy and sent to prison camps, where they received inadequate medical treatment. Some were even taunted if they were disfigured. After being shot twice at the Battle of Mons in 1914, Major Malcolm Vivian Hay was left near the village of Audencourt by his battalion, since there were no stretchers to carry him. He was rescued by a French civilian, who eventually moved him to a hospital in Cambrai, then occupied by the Germans. Four months passed before he was told that he would be transferred to a prison camp at Würzburg. As he boarded a train bound for Germany, he watched as a sentry turned his sights on an Irish prisoner who had been shot in the face. The wounded man's "blind eye was a running sore, the torn cheek in healing had left a hideously scarred hollow, and the mouth and nose were twisted to one side," Hay recalled. The guard pulled the disfigured man out of the pile of prisoners sleeping on the floor while the other sentries stood around, pointing and jeering. Hay wrote that nothing—not even the sight of wounded men being shot—had moved him as much as the "pathetic sight of this young Irishman and his heartless tormentors."

Clare may have felt like a side of beef being carted around, but at least he had not fallen into the hands of the enemy. At the casualty clearing station, he received compassionate care. He was washed, dressed, and given a tetanus shot. Due to the nature of his wounds, a medical officer told him that he would have to be sent back to Britain. There was a limit to what anyone could do for him this close to the battlefield, with casualties streaming in every minute of every day.

Clare was overwhelmed by the thought of returning home. "What the knowledge that I should soon be in England again meant to me then I cannot convey," he later wrote. Not long before, he had been envisioning his own burial. "Having abandoned any hope as I lay wounded on the field in the morning, it seemed all the more wonderful that a few hours should produce such a change." Unfortunately, he would find his hope being tested as he embarked on the long, arduous road to recovery.

From the casualty clearing station, Clare was taken to a base hospital in Rouen, where he was allowed to see his face in a mirror for the first time. "My bristly beard was about half an inch long, and dried blood and dirt in it which could only be removed by shaving, looked so disgusting and so altered my appearance that I was quite unhappy," he recalled. The nurse arranged for a barber to carefully shave Clare, after which he was given a mirror a second time. The extent of his injuries was quite obvious now that the blood and hair had been removed. Shocked, Clare saw for the first time the entrance and exit wounds that had caused so much damage to his face.

A few days later, Clare was loaded onto a hospital ship—one of seventy-seven that were commissioned during the war. The largest in the fleet was RMS *Aquitania*, which had 4,182 beds. On one journey from the Dardanelles, the vessel carried so many wounded men back to Britain that it took twenty ambulance trains to transport them from its berth to various hospitals around the country.

Technically, hospital ships were protected under the Geneva Conventions, since they were not directly involved in combat. However, these floating hospitals—painted bright white and decorated with red crosses—were not immune to danger. Between 1915 and 1917, seven hospital ships struck mines and were either sunk or badly damaged. When HMHS *Anglia* struck a mine just before noon on November 17, 1915, medical staff scrambled to remove the wooden splints from patients' lower limbs, since those who fell into the water with splints still attached would find that their legs floated but their torsos sank. Patients who could walk were marshaled onto deck, while those who couldn't were carried up on stretchers and put onto lifeboats as the ship began to slip beneath the water. In total, 130 of the 388 people on board died that day, including 9 members of the Royal Army Medical Corps.

One of the greatest wartime losses was that of the British hospital ship *Britannic*. She was the sister ship of the infamous RMS *Titanic*, built by the White Star Line just before the war, and she still lies four hundred feet underwater in the Kea Channel off the coast of Greece. Violet Jessop, a nurse aboard the *Britannic* who had also survived the sinking of the *Titanic* in 1912, watched as the mighty ship went down, killing thirty people. "All the deck machinery fell into the sea like a child's toys," she wrote. "Then she took a fearful plunge, her stern rearing hundreds of feet into the air until with a final roar, she disappeared into the depths, the noise of her going resounding through the water with undreamt-of violence." (The uncommonly hapless Jessop had also been aboard RMS *Olympic*, the oldest of the three sister ships, when it collided with HMS *Hawke* in 1911.)

It wasn't only mines that posed a risk to hospital ships. With their fresh coats of white paint gleaming against the blue and gray of ocean and sky, they made easy targets for the German U-boats that patrolled busy waterways and stalked their prey. In 1917, the Central Powers decided to disregard international law. Hospital ships, no matter how well marked, became fair game. During 1917 and 1918, a number of ships carrying injured soldiers were torpedoed. Most tragic of all was the sinking of HMHS *Llandovery Castle*, which was sailing from Halifax to Liverpool when a U-boat attacked it on the night of June 27, 1918. There were no patients on board, but there were numerous nurses. Lifeboats and rafts were deployed, but the German submarine was relentless, shelling and ramming all but one of the emergency craft. Only 24 of the 258 who had been on the ship when it was struck survived.

Percy Clare, at least, made it safely across the Channel—though the journey was tense. After boarding, he was brought down into the bowels of the ship by way of an electric elevator and was placed in a swinging cot supported by ropes anchored to the ceiling. A nurse secured a lifebelt around him—a stark reminder of the dangers that lurked beneath the waves.

Once darkness had fallen, the hospital ship steamed out of the harbor. The crossing was slow due to bad weather and the threat posed by U-boats. In total, it took thirteen hours to reach Southampton. All through the night, an injured man placed in the cot at Clare's feet cried out in agony. He had been shot through the abdomen, and the bullet had punctured his kidney. Despite a nurse's best efforts, the soldier was dead by morning. "I saw tears run down the sister's cheek, and marvelled at the tenderness she felt for a man she did not know," Clare wrote.

The ship finally reached the shores of Britain early the next morning, just as the sun was peeking over the horizon. A swarm of Red Cross volunteers began off-loading casualties. Clare was once again placed on the ground, along with three thousand other injured soldiers. The journalist Philip Gibbs was haunted by a similar scene. "Outside a square brick building," he wrote, "the 'bad' cases were unloaded: men with chunks of steel in their lungs and bowels were vomiting great gobs of blood, men with arms and legs torn from their trunks, men without noses, and their brains throbbing through opened scalps, men without faces."

The label pinned to Clare's uniform stated that he should be sent to a "special London hospital," presumably Sidcup. Unfortunately for Clare, he was mistakenly put onto a train that was heading in the opposite direction. When he was told this, he turned his bruised and shattered face to the window and quietly began to weep.

While Harold Gillies was operating on soldiers at the Queen's Hospital, Clare lay in a facility sixty miles away. The Frensham Hill Military Hospital was established in the moderately sized home of one Mrs. Lewin in October 1914 and was staffed by the Frensham Voluntary Aid Detachment. In time, barracks were erected nearby in order to house the influx of wounded soldiers. But even with these additions, the Frensham Hill Military Hospital was a small outpost compared to other medical facilities at that time. Crucially, it didn't offer the type of specialized care that Clare's facial wound so desperately needed.

"The Hospital at Frensham was a poor sort of place," Clare wrote in his diary, "just temporary wooden structures each with a 'Tortoise slow-combustion' stove in the centre." Chief among the staff was the matron, or head nurse. Clare described her as a "vinegar faced old 'cat'" who wanted the men to stand at attention whenever she entered the room and who punished what she perceived as insubordination by withholding margarine at mealtime. The hospital lacked the jovial atmosphere so many patients enjoyed at Sidcup. Here, there was no piano-playing or card games or secret late-night snacks. "[T]he men had no entertainment or means of passing the time away at all," Clare complained. He described his days there as "deadly dull." Most of the time, he and the other men crowded around the stove for warmth while smoking cigarettes and chatting aimlessly about the war.

Clare faced problems far more serious than boredom, however. He found the food at the hospital largely inedible due to his injured jaw, which prevented him from being able to chew the hard chunks of bread provided with each meal. "There was no provision for such a case as mine," Clare wrote. "My jaw was swollen and stiff and I had no power to open it."

The medical staff didn't know what to do with him. Although Clare's wounds had been cleaned and bandaged, he had yet to receive any specialized care. Unfortunately, delays in treatment would only cause problems down the line, as scar tissue formed and infection set in. One doctor admitted to Clare that he had been sent to the wrong place. "The MO [Medical Officer] wanted me to get up as soon as possible, and get out for walks so that I should get strong enough to travel to a special hospital," Clare wrote. "He said they could not deal with my case . . . [and] that I should never have been sent there."

While he waited to be transferred, Clare discovered that another man from his company had also been sent to the Frensham Hill Military Hospital and was housed in a separate section. Clare sought out his old comrade, who casually confessed to murdering their commanding officer in the midst of battle on the day that both he and Clare had been injured. "He told me boastingly that he had got even with 2nd Lieut. H—s, who had several times had him brought up before the Company Commander for punishment," Clare wrote in his diary. Later, Clare was able to confirm that this particular officer had indeed been shot in the back several times.

That wasn't the only troubling news he received during his stay. Rawson—the officer who had shoved the packet of emergency field dressings into his mouth after he had been shot—wrote to Clare, informing him that most of his comrades had been killed at Cambrai. Not for the first time that wintry month, his mind turned to death. Clare wondered what his own fate would have been had he not been shot in the face just minutes into the advance. "Had I not been wounded . . . should I have been able to endure and survive?" he asked himself.

After languishing at Frensham for weeks, Clare was told that he was finally going to be transferred to the Queen's Hospital. In early December, he was loaded into an ambulance along with an officer who was to act as a chaperone. As they bumped along the gravel roads, the man told him that they would have to go into the city in order to catch a train from London Bridge station to Sidcup. Clare's ears pricked up. He asked that he be allowed to part ways with the officer for a short while so he could visit his wife, Beatrice, for the first time since being injured. She was a customs officer at Kearley & Tonge's warehouse in East London. His companion agreed but warned Clare that he would need to be at London Bridge station by late afternoon if he was to catch the last train to his destination.

"Imagine my feelings at the thought of seeing her whom I had not seen since my 'draft leave' before I went to France in Sept 1916," he wrote. By the time he reached the warehouse, however, he was beginning to feel faint from exhaustion. Worse than this, though, was the news that his wife was off work that day. Dejected and exhausted, Clare made his "very sorrowful way" to London Bridge station. By the time he reached the platform, the train to Sidcup had already departed.

Setbacks had dogged Clare at every turn, but he suddenly experienced a stroke of luck. As he stood there, frustrated and dejected, he heard the click-clack of heels approaching from behind. A group of women had noticed his bruised and bandaged face and wanted to know if they could help. They congregated around him as he explained that he had missed his train and was now stranded.

The women took charge of the situation and led Clare from the station. They were organizers of the "Soldiers' and Sailors' Free Buffet" at Victoria Station, which employed hundreds of female volunteers who worked around the clock in twelve-hour shifts to feed troops moving into and out of the city. Between 1915 and 1919, the organization fed over eight million servicemen in the capital. The women served Clare tea and refreshments while they arranged for an urgent telegram to be delivered to Sidcup. Later that evening, they bade him farewell after securing his passage on another train. "What kind friends the Tommy ever found in those War days," Clare wrote.

>—<

The Queen's Hospital was brimming with holiday cheer. It was a few weeks before Christmas, and the wards were in the middle of a fierce competition to see which one could boast the most festive decorations. A gramophone spun continuously as the men hung large garlands from the rafters. Even the hospital mascot, a parrot, seemed to be in a chipper mood despite the shorter daylight hours.

Clare enjoyed a warm reception when he finally arrived at Sidcup in December 1917. "Men came round the bed to enquire about me," he wrote. "They all seemed just one happy family, each thinking of the other fellow's welfare." He was immediately struck by the differences between Sidcup and Frensham Hill. The dark-green enameled beds were covered in "snowy white" linen that looked warm and welcoming. At the foot of each cot was a scarlet woolen blanket, which was set off by the green-gray walls of each ward. The combination gave the hospital a cheerful look in the dead of winter. A nurse led him to his bed, helped him undress, and gave him a warm glass of milk—a kind act that contrasted with the admonishments of the "vinegar faced old 'cat'" back at the last hospital, who punished the men for insubordination by withholding food.

The nursing staff won the admiration of everyone at the Queen's Hospital. One soldier observed that the nurses "entered into their work so zealously that we are indebted to them as much as anything else, for the peace we now enjoy." The nurses, in turn, did their best to provide the men with top-notch care. Nellie Cryer, who began working at Sidcup shortly after the hospital opened, believed "there could not have been a more appropriate place for those kind [sic] of patients."

Small touches made a big difference. Hours after Clare arrived, a night nurse came onto the ward to hang lamps with red shades at the end of each cot. "I lay staring at the shaded light which threw a soft warm glow all down the centre of the ward and spread over the polished wood floor," he wrote. "I was so happy on that first night in the comfortable cosy ward, beautifully wide, with many beds and bright congenial company, that in spite of my fatigue I couldn't sleep a wink."

Since opening, the Queen's Hospital had grown into a thriving community of patients and practitioners, all of whom felt a deep kinship with one another. The atmosphere could be joyful despite the grim circumstances that brought people there. "They are a cheerful crowd these wounded tommies," Clare wrote in a letter to his mother. His favorite was a young man at the end of the room whose eyes had been "shot out." Somehow, he had managed to hold on to his sense of humor throughout his ordeal, cracking jokes and teasing the nurses at every opportunity.

Unlike Frensham Hill—where soldiers huddled around an inadequate stove and did little more than smoke cigarettes all day to kill time—the men at Sidcup had a variety of ways to occupy themselves. Socializing was made easier by the fact that all the men there suffered from some form of disfigurement. Whereas a man with a facial injury might feel self-conscious about his appearance among men with other types of wounds, he need not feel any embarrassment among the patients at Sidcup. "By removing them from the atmosphere of crowded hospitals, where very often they have shrunk from outdoor exercise and mixing with other patients whose wounds, although painful, are not so obvious," one nursing publication posited, "they will recover completely and heal in a third of the time."

For those who were mobile, there were sports days, during which they could participate in football, cricket, and other athletic activities. Patients could also play croquet or perform in an amateur theater group that staged productions for the amusement of all at the hospital.

Besides leisure activities, the men could attend workshops that would help improve their employment prospects after the war. Some learned how to fix clocks and watches, while others tried their hand at hairdressing and barbering. A soldier could learn how to repair boots and motors or attend courses in bookbinding, photography, and draftsmanship. He could even take up a foreign language—French being one of the more popular options. A visiting journalist admired the "extensive gardens where the patients on approaching the stage of convalescence can be instructed in all manner of outdoor occupations"—such as horticulture, forestry, and poultry farming. The last was especially useful, since trainees could help tend to the countless chickens on the hospital grounds that provided the hundreds of eggs needed to feed the men each day.

One of the more well-attended courses taught soldiers how to make toys. During the holiday season, they produced trinkets that could be sold at various outlets around London. This not only benefited the hospital, but it also instilled in the men a sense of self-worth. Even the royal family delighted in these knickknacks. *The Times* reported that the Queen, along with princesses Mary and Helena Victoria, attended an "exhibition of children's toys . . .

made by the soldier patients of the Queen's Hospital." There was a wide range of beautifully rendered toy animals available for purchase, such as dogs, ducks, monkeys, and camels. One journalist noted with delight that the elephants "had springs cunningly arranged in their legs," which allowed them to bound behind their purchasers "in a manner hardly becoming the elephantine dignity." In the end, the queen chose a small gray chimpanzee to take home to her palace, while a lady-in-waiting to Princess Helena Victoria hugged a "flame-coloured duck" to her chest.

Providing classes to the men while they convalesced was just one of the many ways that the Queen's Hospital stood apart from other medical facilities. Harold Gillies also wanted his patients to feel involved in the recovery process. To that end, he offered them photographic updates on their treatment, so that they might compare their appearance before, during, and after reconstructive surgery. He believed this would keep their spirits up as they underwent multiple, painful operations—though it's unclear whether the patients themselves were encouraged by these photos. Other aspects of the men's welfare were also considered. There was even a hospital barber, trained in special shaving techniques to help tend to faces with deep scars, missing tissue, and with tubed pedicles attached.

It's no wonder that when Clare compared his experiences at the Queen's Hospital to those at other institutions, he declared: "Sidcup was indeed a paradise to me when I arrived."

In a letter to his mother, Clare confessed his greatest fear. "Shall I whisper a secret to you?" he asked. "I'm afraid of getting well. The sooner I recover the sooner 'out there' I go again, and frankly I don't want to go."

If it meant not returning to battle, Clare was more than happy to spend the winter convalescing at the Queen's Hospital while surgeons worked on his face. And there was plenty of work to be done. Shortly after arriving at Sidcup, he met with a dental offi-cer who attempted to pry open his jaw, which had seized up. "He can't do anything until they get my mouth to open," Clare wrote his mother. It was tedious work, but the dental officer eventually loosened Clare's jaw enough for the surgical work to commence.

The first operation that Clare underwent was performed with-out anesthesia. This was likely due to the location and nature of his wound, since it would have been difficult to secure a mask over his face to administer the drugs with a gaping hole in his cheek. It was during this procedure that preliminary work was done to correct any issues that had arisen from the delay to his treatment. The next two operations on Clare's jaw were performed in rapid succession under hallucination-inducing chloroform. "The boys call going under operation in the theatre 'going to the pictures' because of the effects of the anaesthetic," he joked to his mother. "[T]here is much laughter and witty exchanges between the men in the ward-cots and the man on the stretcher, as he is being wheeled out of the ward." Clare was keen to note that while a man might leave laughing, he usually returned moaning. The road to recovery was often a long and painful one.

Clare was not a passive observer while at the Queen's Hospital. In between surgeries, he liked to lend a helping hand. On one oc-casion, he held another man's tongue to prevent him from choking while he recovered from anesthesia. "I had to keep swabbing blood and clots away with my left hand while gently resisting the pull of his tongue with my right." As the man flitted in and out of con-sciousness, he complained to Clare about his inability to smoke due to the large hole in his cheek that prevented him from inhaling.

Over the coming weeks, Clare's jaw continued to seize up. He was instructed to practice opening and closing his mouth through-out the day—a task that proved nearly impossible at times. Still, he made progress, and although it was slow, it seemed that Clare was finally on his way to a successful recovery.

There was to be a boost to his psychological recovery as well. One afternoon, as he was taking advantage of the hospital grounds, the temperature dropped markedly, and snow began to fall. As he took shelter under the frozen branches of the tall elms lining the path to Frognal's main house, he became aware of a familiar figure walking toward him in the distance. Clare's heart raced with anticipation. He had been bitterly disappointed to miss his wife, Beatrice, while traveling through London. But now, after such a long period of separation and so many grave misfortunes, she was standing before him, with snowflakes speckling her hair. "[We] embraced each other closely in the dark under the great Elm trees bordering the road."

Despite the sad circumstances that brought the two together again, it was a happy reunion. Later that evening, as Clare climbed into bed, a feeling of deep contentment washed over him.

Unfortunately, it would not last.

>-•-<

The snow crunched under the soldiers' boots as they were marched out onto the frozen grounds in front of the Queen's Hospital. A warm glow emanated from the windows, reminding those on the outside of the comforts within. It was half past six o'clock in the evening when these men—Percy Clare included—were notified that they would be "Returned to Duty." His greatest fear had become a reality.

Clare was not the first patient to be prematurely discharged and sent back to the front, nor would he be the last. Even when a man had completed his treatment at the Queen's Hospital, a happy ending was not a foregone conclusion. One officer, who underwent a series of painful operations to correct a deep gash that ran from his temple to his chin, was sent back to the front, where he was injured a second time. Gillies noted that the officer "was shot through the knee-joint, and died of wounds in the same casualty clearing station as that which received him when his face was wounded." All too often, the surgical triumphs achieved at Sidcup were effaced by further battlefield tragedies.

The order to return to duty came as a shock to Clare. He still couldn't open his jaw very wide despite enduring multiple operations, and he didn't feel well enough to return to active service. The order, however, had not come from Gillies. Like so many unfathomable decisions made during the First World War, it had been passed down by unseen hands at the top. Clare was told that he was being released to make room for new patients and to ensure that the war machine continued to receive a supply of human fuel. Even in January 1918, there was no end to the conflict in sight. Although Clare felt it was "a blot on England's honour" to turn men like him out before he had completely healed, he knew that he had little say in the matter.

And so, Clare—bruised, swollen, and half-mended—steeled himself for a return to battle.

⇥ 11 ⇤

HEROIC FAILURES

The convalescing officers watched as the pilot's "fleshless fingers" moved gracefully over the keys of the piano. With their faces bandaged, they relaxed in the sitting room of Frognal House at the heart of the estate, sipping whisky through straws and listening to the pianist's lilting tune. The pilot, whose hands had been ravaged by flames after he crashed his aircraft, played beautifully despite the severity of his wounds. Shortly after he had arrived at Sidcup, he had fallen in love with his nurse and married her with the sense of urgency that near-death experiences often provoke. After playing the final few bars of a gentle melody, the pilot suddenly broke out in jovial song: "And now I've got a mother-in-law, [t]hrough drinking whisky through a straw . . ." This elicited a few chuckles and lopsided grins from the wobbly jaws in the room.

Away from the cheerful atmosphere of the sitting room, Harold Gillies sat in his office. It was a cold day in February 1918, and he was turning over in his mind the best way to deal with the miserable condition of Second Lieutenant Henry Ralph Lumley. Another member of the Royal Flying Corps, Lumley had also paid dearly for his willingness to serve.

Powered flight was still in its infancy at the start of World War I. Most planes were used only for surveillance, though some pilots did carry weapons with them, such as guns and grenades, which usually proved ineffective, not to mention downright dangerous. Pilots could fly over enemy trenches to gather information, then drop messages to their own comrades using weighted bags, but these reconnaissance missions were highly risky.

Late in the war, Captain "Freddie" West was flying on just such an assignment. He was far over enemy lines when he was attacked by seven German aircraft. One of his legs was injured, and an explosive bullet partially severed the other, causing it to jam the control pedals at his feet. After dislodging the limb, he maneuvered his aircraft so that Lieutenant William Haslam could fire at the enemy and drive them away. West then twisted his trouser leg into a makeshift tourniquet to stem the bleeding, flew back behind his own lines with the required intelligence, and landed safely. He fainted soon afterward, but when he came around, he insisted on writing his report. He was awarded the Victoria Cross and was eventually fitted with an artificial leg.

West later recalled that the freedom of flying an aircraft through French skies held great appeal to men who were living in cramped, filthy conditions in the trenches. Henry Ralph Lumley was one of those young men who volunteered to do his bit from behind the controls of a biplane. One summer afternoon, Lumley crawled into his single-seat B.E.12 aircraft. It was graduation day at the Central Flying School in Upavon, and he was eager to perform his first solo flight. What should have been a celebratory occasion quickly turned into a tragedy. Not long into his flight, Lumley's plane was crippled by a catastrophic mechanical failure. After losing height over the chalky expanse of Salisbury Plain, he crash-landed the rickety aircraft into one of its fields. The fuel tank, which was in the front of the aircraft, exploded on impact, and the plane was quickly consumed by flames.

Lumley suffered severe burns that destroyed all the skin and most of the subcutaneous tissue of his face. He also sustained burns on his legs, arms, and hands. He was first sent to a military hospital in Tidworth, where doctors removed his left eye. Eventually, he was transferred to King Edward VII's Hospital in the center of London. This had been established by sisters Agnes and Fanny Keyser during the Boer War at the beginning of the century. Agnes was especially moved by Lumley's case. In a letter, she wrote, "His face is burned beyond recognition. One eye removed, the other practically blind." Yet there was a limit to the help doctors could offer him at King Edward VII's Hospital. He was eventually sent home, where he slid into a deep depression.

Try though she did, Agnes Keyser could not forget Lumley. Thanks to her persistence, she eventually secured him a transfer to Sidcup. By then, his injuries had received no surgical attention for over a year, and he had developed deep scars all over his face as a result. After some consideration, Gillies decided that he would need to replace all the skin on the pilot's face using a chest flap— as he had done with the sailor William Vicarage, who had suffered extensive burns at the Battle of Jutland.

Gillies carried out an initial operation to prepare the way for the chest flap. Around this time, he also noticed that Lumley had developed an addiction to morphine. The pilot's general health was deteriorating, and fast. Gillies wrote, "it had to be decided whether to give this unfortunate airman a further year's rest or whether to carry on with the procedure, knowing that the latter might not succeed." Gillies worried that Lumley would simply not be strong enough to withstand another major operation. When he told him this, Lumley was "bitterly disappointed and exceedingly depressed at the thought of having to wait another long period." Gillies decided to proceed, against his better judgment.

Now, an hour before he was due in the operating room, Gillies sat hunched over his desk with a cigarette clamped between his lips, worrying that he was making a grave misstep. It was one of twenty-five cigarettes that he smoked every day. Later in life, Gillies was forced to give up the habit due to his deteriorating health. "It has been a good puff for 49 years . . . it might be said I smoked my way from Dover to Calais or five times around Hyde Park," he would later joke.

Before every major operation, Gillies took refuge in the "tiny narrow room" that was his office in the old mansion at the heart of the estate—away from the bustle of the wards. There, he sketched diagrams of flaps, pedicles, and grafts on a small writing pad, sometimes cutting the designs out with a pair of scissors that he kept in a drawer. Then he carefully laid the pieces out on his desk before fitting them together like a jigsaw. "Can our general surgeons and our doctor friends realise the ever frightening responsibility of that plan," he once asked, "and the irrevocable first cut?"

Gillies ran through the operation in his mind again and again and again, obsessing over every detail and trying to anticipate any problems that might arise. For every patient, he devised what he called a "lifeboat," which was a backup plan in the form of another flap or skin graft. Gillies knew from experience that even the best-laid plans could go awry once he began reconstructing a face. "It is impossible at times to be sure that a flap will fit or look well or even survive," he confessed. "Having made all the plans conceivable for a case, it often happens that at operation the actual plan adopted is a different one."

When Gillies could delay no longer, he put down his sketches, stubbed out his cigarette, and solemnly headed to the operating room. Lumley was already there when he arrived. After the pilot was anesthetized, Gillies picked up a knife and prepared to cut into the chest in order to raise the skin flap. The operation was complex and took several hours. Once the flap was raised, Gillies took a skin graft that had been harvested from a volunteer by Lieutenant Colonel Henry Simpson Newland—head of the Australian section—and transplanted it onto Lumley's raw chest. Surgeons had been grafting donor skin onto patients with varying degrees of success for centuries. The most notable case was performed by the New York surgeon John Harvey Girdner, who successfully performed a skin graft transplant from a deceased donor in 1880. This type of graft, however, carried with it many risks. Gillies later described the process as "exceedingly tedious." But the real problems arose after the last stitches were put in place. It then became clear that Lumley's fragile body could not handle yet another assault.

Gillies reported that the very next day "the patient was considerably collapsed, and the flap itself suffered [from] general depression of circulation, and in thirty-six hours became blue." Soon after, Lumley's surgical wounds turned gangrenous, and the chest graft failed to take. Despite around-the-clock care, the pilot's condition deteriorated rapidly over the coming weeks. "Both the chest area and that of the denuded face became infected," Gillies recorded, "and towards the end metastatic abscesses occurred in various regions." On March 11, 1918, Henry Ralph Lumley's heart gave out.

Despite all the lessons he had learned and all the innovations he had made, failure was Gillies's constant and unwelcome companion at the Queen's Hospital. The death of a patient was just as hard a blow at this late stage of the war as it had been in the earliest days, and he was devastated by the pilot's demise. He blamed himself, later admitting that he felt his desire "to obtain a perfect result" overrode his surgical judgment. Instead of trying to reconstruct Lumley's entire face at once, he believed he should have undertaken the work piecemeal, addressing one quarter of the face on each occasion. He was plagued by questions, wondering how the outcome might have differed if he had "taken a very firm attitude" with Lumley and convinced him to delay his operation. With a heavy heart, Gillies confessed, "One could have wished that this brave fellow had had a happier death."

Lumley's passing came just days after Russia withdrew from the war by signing a peace treaty with the Central Powers. Russia's decision—prompted in part by the overthrow of Tsar Nicholas II in March 1917 and the Bolshevik Revolution eight months later—effectively ended fighting on the Eastern Front. This meant that the Allies on the Western Front would soon be faced with hundreds of thousands of additional German troops. The British officer Richard Tobin recalled a sense of foreboding: "In the trenches at night, when the wind was in the right direction, we could hear the German trains and transport rumbling up their great army that was going to sweep us into the sea. We were grim, we were determined. Behind us lay the old Somme battlefields, every yard soaked with British blood shed through almost two years of hard battle."

Global setbacks such as these—and those closer to home, such as the loss of stoic patients—must have given Gillies the sense that he was waging his own war on more than one front. But while experience was a brutal teacher, its lessons were well understood. In the aftermath of Lumley's death, Gillies began to adopt a more incremental approach to reconstructive surgery. As he learned with Lumley, each patient's individual needs had to be taken into consideration when formulating an operative plan. What had worked for Vicarage had not worked for Lumley and might not work for future patients. "Never let routine methods become your master," he warned. The setback validated Gillies's philosophy of putting off today what could be done tomorrow.

He would not make the same mistake again.

>—•—<

Henry Tonks loomed over Daryl Lindsay's shoulder. The Australian artist—a "biggish man" with powerful shoulders and a smashed nose that gave his face a "misleading air of pugnacity"—

was hunched over his easel when Tonks's shadow darkened his canvas.

Lindsay had been working as an assistant to the war artist Will Dyson when he was called upon to draw medical diagrams for Lieutenant Colonel Henry Simpson Newland, the surgeon in charge of the Australian section at Sidcup. "I was due back in France the next day, but [the commanding officers] fixed up an extension of leave," he recalled. Only later did Lindsay discover that the paperwork detailing his transfer had not reached the appropriate authorities back in France, and he had been listed as A.W.L. (Absence Without Leave) for thirty days—an offense punishable by death.

Like so many people working at the Queen's Hospital, Lindsay came to be there by sheer happenstance but was quickly thrown into the fray. On his first day, he bumped into Newland, who was making his way to the operating room. "I can see him now, standing with his gloved hands clasped together waiting for the patient to come in from the anaesthetic room," Lindsay later wrote.

Lindsay introduced himself to the surgeon, who invited him to witness the next procedure. "He explained to me that he was going to do a second stage of a rhinoplasty for the restoration of the nose; he said I could leave if it upset me." As it turned out, Lindsay was inured to the sight of blood and gore, having spent considerable time at the front. But he had a more pressing concern as he watched Newland operate. "How was I going to translate what looked like a mess of flesh and blood into a diagram that a student could understand?" he wondered.

Afterward, Lindsay met with Newland for lunch. He told the surgeon that the officers who had arranged for his transfer to Sidcup had "sold him a pup." He confessed that he knew nothing about anatomy and was doubtful he was qualified for the job. Newland smiled gently. Who here didn't feel underqualified to take on the monumental task of rebuilding men's faces? He asked Lindsay to give it a try before quitting. Lindsay reluctantly agreed, and it wasn't long before he began to delight in his role as artist for the hospital's Australian unit.

Now, as Lindsay fussed over a portrait of a patient, Tonks considered the younger artist before him.

"What are you doing?" he asked.

"Trying to draw," Lindsay replied, only paying the man behind him cursory attention.

"I'm glad you said 'trying,' which is the best that can be said of it," Tonks shot back. Lindsay must have looked crestfallen, because Tonks quickly added, "I think I may be able to help you."

And so it transpired that Daryl Lindsay came to spend one day each week at the Slade School of Fine Art in London under the tutelage of Henry Tonks. "Tonks, with his piercing hawk-like eye, was an intimidating person," Lindsey wrote, "and the students of the Slade were terrified of him." Unlike his peers, however, Lindsay would not be cowed. His irrepressible spirit earned him a dinner invitation from Tonks on more than one occasion. The meal was always intimate, with no more than four guests in attendance, which Tonks considered to be the perfect number for good conversation. "He demanded the best and would not tolerate anything second-rate," Lindsay observed. In time, Lindsay's medical portraits won the approval, if not the outright admiration, of the great Henry Tonks. The pair became lifelong friends.

Early on, Harold Gillies had recognized the importance of documenting his work so that others might learn from it. "Not a small feature in the development of [plastic surgery] is the compilation of records," he wrote. This would not have been possible without the help of the artists at Sidcup. The portraits provided a visual record of the cases from start to finish, while surgical diagrams helped others replicate the complex procedures that restored form and function to the soldiers' faces. Over time, the artists at the Queen's Hospital became crucial members of the reconstructive team. Harmony between the creative and medical disciplines was both unique and essential to the practice of plastic surgery.

Besides artists like Lindsay and Tonks, there were sculptors like Kathleen Scott. Tonks had met Scott in France back in 1915, when she led a small ambulance service near the Western Front, and the two were happily reunited when the Queen's Hospital opened. While studying in Paris, she had become friends with Auguste Rodin—the world-renowned artist who led the way for modern sculpture. Rodin's influence can be seen in the fluidity of Scott's early works. She had famously created sculptures of her late husband, the ill-fated Antarctic explorer Robert Falcon Scott, as well as of Edward Smith, the captain of the *Titanic*. While at Sidcup, she created plaster casts of the faces of men with less celebrated but equally heroic stories.

There were also photographers at the Queen's Hospital. Chief among them was Sidney Walbridge, who had first encountered Gillies at Aldershot, where he had been stationed in December 1916. When Gillies moved to Sidcup, Walbridge went with him. Photography was a quick and efficient way for surgeons to document each case. Walbridge directed the patient into a chair with a headrest and took photos from as many as five different angles. The precision of the posing allowed for exact comparisons at various stages of the reconstructive process. In time, these photographs provided another historical record of the birth of modern plastic surgery.

Of course, Gillies was not the first medical professional to use photographs to document cases. The earliest medical photograph of a disfigured patient dates back to 1848. It depicts a burn victim's distorted face and neck. And a few years later, some of the earliest pre- and post-operative photographs were made during the American Civil War. By the late nineteenth century, many doctors believed that the lens of a camera was a powerful tool for achieving objectivity. As a result, the medical community embraced photography as a technology with great potential, especially as photos could be taken with relative ease and at a low cost. Only later would the ethics of such images be challenged.

Despite the presence of photographers, sculptors, and artists at the Queen's Hospital, Gillies continued to dabble in painting. He was a competent rather than exceptional artist, though he fancied himself more talented than he was. One day, Gillies proudly displayed two of his paintings to Gay Tydeman, a medical illustrator who had studied under Tonks. It didn't take a minute for her to dismiss him as a "fairly average photographic-type painter." Gillies left the room in a sulk, carrying one picture in each hand. As he departed, an orderly turned to Tydeman and said, "You didn't leave 'im much 'ope, Miss, did you?"

Tydeman may not have rated Gillies highly as an artist, but she was in awe of his surgical skills. "[H]e was as full of ideas as a dog of fleas," she recalled. "If they succeeded, they were magnificent. If they failed, it was on an heroic scale."

Gillies himself knew this all too well.

The trees outside Frognal House were garlanding themselves in spring blooms. But the promise of a kinder season seemed worthless alongside the developing threat on Europe's battlefields. With troops newly arrived from the Eastern Front, the Germans were able to achieve some startling gains in the spring of 1918, inflicting heavy casualties on the Allies and ending years of stalemate. One British soldier remembered thinking, "Oh God, this is the end," as he watched the massed ranks of Germans break through in formation. It was beginning to look as if they might win a stunning victory after all. Only the United States had the power to shift the balance. Since entering the war in April 1917, the country had been conscripting and training hundreds of thousands of troops.

But deployment had been relatively slow, and on May 2, General John J. Pershing—commander of American forces in Europe—agreed to ramp up pressure on the Germans by sending tens of thousands of fresh troops to fight alongside the French and British forces. Hope was on the horizon for the Allies.

Gillies leaned back in his chair and took a drag from his cigarette, eyeing a letter on his desk. It was from his old colleague Auguste Charles Valadier, the French dentist who had transformed his Rolls-Royce into a dental workshop and driven it to the front under a hail of bullets. "Private Bell is a very fine chap and deserves your personal attention," Valadier had written. This was not the first time his old colleague had contacted him, nor would it be the last. But this particular letter worried Gillies.

After Gillies returned to Britain in 1915, Valadier had continued his work at the specialist unit in France. He toiled around the clock, experimenting with bone grafts and other innovative techniques to offset tissue loss. Over time, however, Valadier realized that there were limits to what he could achieve so close to the front and with resources in such short supply. He also faced professional barriers, since his dental qualifications did not allow him to operate without medical oversight. During the final years of the war, the authorities trimmed back Valadier's duties to such an extent that his unit had become little more than a clearing station for facial injuries. When given a choice, Valadier preferred that his cases be transferred to Gillies's hospital, if they had to be transferred elsewhere at all.

One such transfer was Philip Thorpe of the King's Liverpool Regiment, who was hit by a shell that severed most of his lower lip and a large portion of his jaw. Before Thorpe was shipped off, Valadier had wired the two ends of the mandible together and attached an expansion screw to a vulcanite plate. This was then used to push the fractured ends apart, slowly and incrementally, in order to stimulate new bone formation. As successful as this treatment was, there came a point at which Valadier had done as much as he could. So he sent Thorpe to Sidcup, where he was first operated on by the Canadian division. After several botched efforts, Thorpe grew frustrated and asked to be discharged. It was then that he met Harold Gillies. "[H]e offered to do the job himself, and guaranteed that one operation would finish it," Thorpe later recalled. "He kept his word."

Valadier's most recent referral was the Private James Bell mentioned in his letter to Gillies, which detailed the young man's harrowing case. Before Bell had arrived at Sidcup, he had been treated at the 83rd General Hospital, where surgeons hastily stitched together the deep gash in his face without first addressing the extensive tissue loss he had sustained on the battlefield. Yet again, Gillies was exasperated by the early closure of a wound that compromised the integrity of the face's underlying structure.

Worse still was the fact that Bell's upper lip and nose had been severely damaged and were semigangrenous. Despite doctors' best efforts, the flesh around his mouth sloughed away, taking with it what was left of his lips. By the time Bell had reached the Queen's Hospital, his face was a complete mess. "His little mouth opened vertically and the vicious position of his nose could be seen in a glance at his profile," Gillies recorded in his casebooks. Here was yet another wounded man who would have benefited from careful planning before undergoing reconstructive surgery.

Gillies knew the task ahead wouldn't be easy. In order to reconstruct Bell's face, he would have to undo previous surgical mistakes, which meant the young man would look worse before there was a chance he would ever look better. "Obviously the patient could not remain in this condition," Gillies wrote, "but it was not without fear that I began to undo all that had been done to him." He would never dream of betraying any such anxiety to his patient, however.

As he had done on many occasions, Gillies sequestered himself in his office ahead of the operation. Valadier's letter lay nearby as he rehearsed his plan for Bell's face over and over again in his mind. He was most concerned with the soldier's nose, which had miraculously survived despite the infection. Nevertheless, he knew that one small error could result in its destruction, given its fragile state. Gillies consulted his notes and sketches as the minutes ticked away. When he could procrastinate no longer, he rose from his desk and headed out of the stately home that was once the heart of the Frognal estate. He strode across the pristine lawn toward the newly built compound where his patients resided and in which he had invested so much of himself.

Sunlight filled the operating room as Gillies entered. He went straight to the basin, where he began the preoperative ritual of vigorously scrubbing his hands and forearms before donning a pair of surgical gloves. A nurse presented him with a linen bag containing a sterilized gown wrapped in muslin. Gillies carefully took the garment by the neckband and put it on, while an attendant tied the strings. Bell was already there, surrounded by several other members of Gillies's team. The soldier's eyelids were beginning to droop from the drugs being administered to him by the anesthetist.

After Bell was sedated, Gillies chose a scalpel as carefully as he might choose a golf club. He paused briefly before making that "irrevocable first cut." As Gillies began excising the dense, contracted scar tissue around Bell's mouth, the corners of the soldier's lips sprang back into their normal position. Bell's nose, however, took on a "horrid blue colour." Gillies started to sweat as he continued working at the scar tissue until the nose gradually shifted back to the center of Bell's face and regained its natural color. With the work concluded, Bell was wheeled through the hallways of the makeshift hospital and back onto the ward.

As Gillies had predicted, Bell looked worse for having undergone the operation—though he wouldn't have known this, since he was swaddled in layers of bandages and denied access to a mirror.

Bell's features were swollen beyond recognition, but Gillies wasn't concerned. He could see beyond the angry swelling to what Bell would look like once the reconstructive process was complete. Gillies was excited by the result, since he felt that it underlined the value of a fundamental principle of plastic surgery. "The first step toward filling a tissue gap was to keep what was normal in its normal position," he wrote, "or [as in Private Bell's case] . . . move it back into its original normal position and retain it there." This, he believed, was the cornerstone of this strange new art.

After the initial operation, the dental team was able to replace missing bone in Bell's upper jaw with a vulcanite prosthesis fitted with porcelain teeth. This provided a natural contour that Gillies then used to construct a new top lip. He did this by taking a series of skin flaps from Bell's cheeks and chin to create both the outer surface and the inner lining of the lip. Following a long series of operations and even longer periods of recovery, Gillies could finally write, "I was more than thankful for its satisfactory result."

Bell's case was a difficult one, given the severe loss of tissue exacerbated by the hasty closure of his primary wounds at Valadier's hospital. The reconstructive work required specialist skills that most surgeons at the time simply did not possess. Stitching a large cut on a leg was nothing compared to the delicate task of sewing together a deep cut to the face. "A good style will get you through," wrote Gillies. "Surgical style is the expression of personality and training exhibited by the movements of the fingers; its hallmark—dexterity and gentleness." As Harold Gillies illustrated time and again at Sidcup, the plastic surgeon was more than just a competent craftsman. He was, above all else, an artist.

❈ 12 ❈

AGAINST ALL ODDS

apers of sunlight pushed through the tall windows of the operating room, seeming to set ablaze everything they touched. Despite the brightness, Gillies felt a sudden drowsiness crash over him. His scalpel hovered uncertainly over his patient as his leaden eyelids began to droop. Ether was escaping into the air through the patient's own exhalations and was now threatening to anesthetize anyone nearby. Gillies bore the brunt of it as he bent over the soldier's face.

While the challenges facing Gillies in the operating room were colossal, there were even greater challenges facing Sidcup's anesthetists. Like many aspects of medicine, anesthesia was poorly understood at the time of the First World War. Its practice had changed little since the mid-nineteenth century, when ether's anesthetic properties were first discovered. Anesthesia as a subspecialty did not yet exist. This meant that anesthetics were often administered by a junior doctor, at least at the start of the war, rather than by a specialist who understood the effects of certain anesthetic agents on the severely injured. Indeed, anesthesia had only become a part of the medical curriculum in Britain in 1912.

Unsurprisingly, there was a great need for anesthesia at the front. Over the course of the war, the British alone used 413,198 pounds of ether and 249,341 pounds of chloroform, not to mention hundreds of thousands of gallons of nitrous oxide. The need was so great that nonmedical personnel were sometimes enlisted to anesthetize patients. The Reverend Leonard Pearson recalled performing this duty at the 44th Casualty Clearing Station during the Battle of the Somme:

> I spent most of my time giving anaesthetics. I had no right to be doing this, of course, but we were simply so rushed. We couldn't get the wounded into the hospital quickly enough, and the journey from the battlefield was terrible for those poor lads. It was a question of operating as quickly as possible. If they had had to wait their turn in the normal way, until the surgeon was able to perform an operation with another doctor giving the anaesthetic, it would have been too late for many of them. As it was, many died.

The sheer volume of patients requiring anesthesia, however, was just one of the many problems facing medical personnel. And there were additional challenges when it came to sedating patients with facial wounds.

The conventional method for administering ether or chloroform, which involved placing a gauze mask over the face, often obscured the surgical field. Even when a single rubber tube was passed through the nose or the mouth to administer the vaporized ether or chloroform by hand bellows, the surgeon and anesthetist could find themselves getting in each other's way, as both needed direct access to the face. As a consequence, putting the patient under could be a logistical nightmare in the operating room. Captain Rubens Wade, who worked as an anesthetist alongside Gillies at both Aldershot and Sidcup, wrote that "the surgeon must perforce trespass upon the territory usually regarded by the anaesthetist as his own."

The drugs themselves were also problematic, as they often induced extreme nausea in the patient—a situation that was far from ideal for someone with a severe facial injury. "[W]hen a boy was notified of an operation for the following Monday, he began vomiting on Saturday," Gillies joked. Patients were often more fearful of the anesthesia than the surgery itself: "the sight of a man in a white coat hovering near with a chloroform bottle and gauze pad in one hand, a tongue forceps in the other, often terrified patients of a generation brought up in dread of the surgeon's knife." Many soldiers were also heavy smokers, which made it difficult to anesthetize them with ether or chloroform, since nicotine can affect the way the body metabolizes certain drugs. Some were suffering from chronic bronchitis or other upper respiratory conditions that presented yet more complications.

One of the greatest challenges by far, however, was posed by the number of blood vessels in the face. If a patient's blood pressure was too high, he would bleed excessively. Not only would the blood then obscure the area requiring attention, but it might also drip back down the throat and into the lungs, causing the patient to drown in his own fluids. One solution was to sit a patient upright. But this too presented challenges. "Positive pressure was necessary to prevent blood from entering the trachea," remarked Ivan Magill, an anesthetist at Sidcup. "[B]ut the surgeon got the blast of a patient's ether-laden expirations and was often enveloped in a spray of blood." Gillies was frequently on the receiving end of this unpleasant effect.

Even this late into the war, medicine was still struggling to address the bewildering variety of damage that modern weapons could inflict upon the human body. And not all the problems that surgeons faced would be resolved before the conflict ended. In 1919, Magill and his team would improve methods for administering anesthetics by using a motor pump to push vaporized ether through a catheter placed in the patient's trachea. Endotracheal insufflation, as it's now known, reduced the chances of anesthetic shock by allowing the anesthetist better control over the quantity of drugs entering the patient. Magill would eventually add a second tube to his system—one to deliver anesthesia, the other to prevent ether and blood-laden expirations from hitting the surgeon on their way out of the patient. Just as Gillies promoted the cause of plastic surgery after the war, Magill would later advocate for the establishment of anesthesia as a specialty in its own right, and he would become one of the most important figures in his field in the twentieth century.

But for now, those who made their way to the Queen's Hospital would have to wait a little longer to benefit from such advances.

War and its consequences were undoubtedly driving medical innovations, and Gillies's team frequently put new methods to good use. But there were never any guarantees of success at the Queen's Hospital.

Private Stanley Girling, who sustained serious injuries while fighting with the 72nd Seaforth Highlanders in France, was transferred to Sidcup shortly after being wounded—presumably due to the severity of his facial injuries. By the time he arrived, however, the amount of blood he had lost was of grave concern. Although emergencies involving unchecked bleeding were not something Gillies had to deal with often, any improvements in blood-transfusion techniques were in the plastic surgeon's interest, since the tissue of the face is so vascular. But just as anesthesia had yet to be perfected at the outbreak of World War I, transfusions were also rarely performed due to the high risks associated with them. Finding a safer and more effective method of blood transfusion would become a grave necessity for doctors treating soldiers at the front.

The first recorded blood transfusion took place in 1666, when the English physician Richard Lower transferred blood from one dog to another. Attempts to transfuse blood from animals to humans followed, leading to numerous fatalities, accusations of working against nature, and fear of grotesque side effects, such as recipients sprouting horns. As a result, the practice was largely abandoned.

It wasn't until the nineteenth century that the first human-to-human transfusions were trialed. Between 1818 and 1829, the Englishman James Blundell performed a series of transfusions in which fewer than half his subjects survived. Blundell was at a loss to explain this. In the early twentieth century, the puzzle was solved by the Austrian physician Karl Landsteiner.

For decades, doctors had noticed that when blood from different donors was mixed, the cells sometimes clumped together. Since the blood in question often came from sick patients, most doctors considered this an abnormality unworthy of investigation. Landsteiner, however, wondered how blood from two healthy people would interact. So, he collected blood from himself and his colleagues and found that clumping of cells only occurred when certain people's blood was mixed—regardless of the health of the donors. He sorted his samples into three groups labeled A, B, and C (the last was eventually renamed O after the discovery of a fourth group, AB). When he mixed like with like, the blood remained liquid. But on mixing A and B together, the cells clumped. Further, mixing A or B with C (or O) did not result in clumping.

Landsteiner realized the immune system was responsible. Blood contains antigens, which cause the body to produce antibodies to fight off invaders, such as viruses. Each blood type has different kinds of antigens. When different types mix, the immune system attacks the foreign antigens, causing the blood cells to agglutinate. When this happens, the recipient develops blood clots, which can be fatal. The exception is type O, which has no antigens and is therefore compatible with the three other blood types.

But even crossmatching did not suddenly make blood transfusions safe or easy. Surgeons still had to cut through the skin to expose the blood vessels and then connect the donor and recipient with a rubber tube in a method known as direct transfusion. Both people had to lie perfectly still side-by-side for hours so as not to break the connection, and it was nearly impossible to measure how much blood actually passed between them.

In 1913, New Yorker Edward Lindeman devised a less invasive method. First, he inserted a hollow tube, or cannula, into the donor's vein, which he then attached to a glass syringe. This allowed him to withdraw a measured amount of the donor's blood. Once the syringe was full, he removed it and transferred the blood to the recipient, who had a similar cannula inserted. This method allowed precise amounts of blood to be transfused between two people. However, the process had to be highly choreographed, as delays could result in blood clotting in the syringes. But a further development on the eve of the First World War addressed this problem.

The Belgian doctor Adolf Hustin discovered that sodium citrate acts as an anticoagulant when mixed with blood, allowing it to be stored for later transfusion. In March 1914—just four months before the start of the war—Hustin performed the first transfusion of citrated blood in Brussels. "This great stride forward in the technique of blood transfusion coincided so nearly with the beginning of the war that it seemed almost as if foreknowledge of the necessity for it in treating war wounds had stimulated research," the British surgeon Geoffrey Keynes later wrote. But storing blood at the front was not yet feasible, so doctors often performed direct infusions—if they performed them at all.

The first soldier to receive a transfusion during the war was twenty-five-year-old Henri Legrain of the French army's 45th Infantry Regiment. After being injured in the trenches near Maricourt during a day of heavy bombardment, the young corporal was transferred to a converted hospital at the Hôtel du Palais in Biarritz. He had already lost a tremendous amount of blood, and there were no signs that the bleeding would stop anytime soon. Lying in an adjacent bed was Private Isidore Colas, who was recovering after his leg had been ripped apart by shrapnel. When Emile Jeanbrau—one of the attending doctors at the hospital—asked Colas to donate blood, he readily agreed. On October 14, 1914, Colas and Legrain were connected by a silver tube that allowed blood to pass between them for close to two hours. Little by little, color returned to Legrain's face. When the procedure was over, he felt so much better that he leaned over and kissed Colas on both cheeks. Legrain had been incredibly lucky, since the doctor did not have the time or resources to crossmatch blood types before the transfusion.

Isolated success stories notwithstanding, blood transfusions at the beginning of the war remained few and far between—due in part to high failure rates. In 1916, a surgeon named Andrew Fullerton performed indirect blood transfusions on nineteen injured soldiers at a casualty clearing station in Boulogne. He collected donor blood in paraffin-lined tubes to prevent it from clotting outside the body and then transported it to an adjacent room, where he transfused the blood into the recipients. Despite Fullerton's heroic efforts, however, fifteen men died. The truth was that the existing techniques (whether direct or indirect) were too difficult and time-consuming and often required a specialized team of surgeons, of whom there was a shortage on the front. Even so, Fullerton maintained that blood transfusions "ought to be used much more widely than is the case at present," albeit in only the most desperate situations, given the high risks.

It wasn't until 1917 that further technical advances enabled easier blood transfusions. Crucial to these advances was Oswald Hope Robertson, an American hematologist from the Rockefeller Institute Hospital in New York. Robertson was sent to Base Hospital No. 5 in France, which was under the direction of the famed neurosurgeon Harvey Cushing. The young hematologist had very little clinical experience and was initially overwhelmed by the sheer number of casualties. "Being pitched into a hospital's service of 100 beds with little in the way of conveniences to work with, and the main purpose in view to get rid of your cases as fast as possible, is somewhat disturbing after the peaceful serenity of the laboratory," he wrote in a letter to a colleague back home.

What struck Robertson most was the challenges doctors faced when transfusing blood. He opined that the "difficulty of procuring sufficient blood under rushed conditions, the time consumed in carrying out the transfusions, and the need of every available medical officer in the operating theatre all tend to reduce the number of transfusions which can be given." That was when he began to wonder whether it would be possible to keep "blood on tap." To this end, he designed his own apparatus for transfusing blood and began thinking of ways to store it in advance of its being needed.

In November 1917, Robertson was transferred to a casualty clearing station near the Western Front in preparation for the Battle of Cambrai (at which Percy Clare would be wounded). Before departing, he packed glass jars of citrated blood from universal donors in an ice-filled chest that he had constructed from two ammunition cases. When he arrived, he began performing transfusions using the stored blood. Unfortunately, he ran out on the third day of battle and had to resort to alternative methods. He later wrote that "[it] was then that I realized what a tremendous advantage the preserved blood was," adding that his technique was "quite the show during the push." Indeed, the lives of many men were saved during the Battle of Cambrai due to Robertson's improvised cache of blood.

Confident in the value of blood transfusions and the need for donors, he spurred his medical colleagues into action. He was able to convince a colonel to donate his own blood during one of his many lectures on the subject. "Banking," as it became known, allowed blood to be collected in advance from preselected donors to supply the needs of frontline recipients. The British Army even began granting extra leave to soldiers with the universal blood type who donated their blood. It also established resuscitation teams to carry out transfusions before and after surgeries. A single doctor, usually accompanied by one assistant, could transfuse blood quickly at the patient's bedside without having to move the recipient and donor into an operating room. This was not only easier, but it also freed up space in the operating room for other procedures. One resuscitation team was led by the Scottish physician Alexander Fleming, who would later discover penicillin.

Despite these positive changes, Robertson was ribbed by some of his colleagues for his obsessive focus on blood transfusions. "To the serious scientist the war has been very, very bloody," one critic wrote. "When the carnage became insufficient to satisfy his curiosity, he grouped together about him a collection of volunteers and drew upon them to his heart's content. No leech of old ever applied himself to a subject more firmly or got so much out of it in the long run." Teasing aside, the advances made by Robertson were undeniable. It wasn't a perfect system, but it was far better than any existing at the start of the war.

As a result, when Private Stanley Girling was taken in by Gillies's team at the Queen's Hospital with severe blood loss after incurring injuries to his left shoulder and jaw, he had a much better chance of survival than the soldiers who had come before him. The totality of what doctors could then achieve with blood transfusions was far greater than it had been at any point in history. Moreover, Girling's odds were improved by the fact that he had an older brother, Leonard, who was working at the Royal Arsenal in Woolwich, not far from Sidcup. There was no need to find a donor among the injured men crowding the hospital wards when there was a viable donor in Stanley's brother—a strong, healthy man with no known medical conditions and who was likely a familial blood match.

When Leonard was told that his little brother had been seriously injured, he rushed to Sidcup to be at his side. There, doctors asked him whether he would be willing to donate some of his blood. He did not hesitate. Within hours, Leonard lay on a hospital bed next to Stanley while blood from his veins flowed into his brother's. Unfortunately, the first transfusion was insufficient, and so a second was performed. Stanley slowly regained this strength. But his restored health came at great expense, for his brother—once hale and hearty—did not fare as well. Whether it was from blood loss or something else, Leonard collapsed a day later and, quite unexpectedly, died. Newspapers declared that the young man had made a "splendid sacrifice" to save the life of his brother.

No mention was made of Stanley's feelings about his loss.

>-•-<

By July 1918, Henry Tonks had seen everything: jaws ground to pulp, eyeballs dislodged, noses replaced by deep craters. He had experienced personal loss too. Recently, he had been notified that one of his former students at the Slade School of Fine Art had been killed in battle. But he would witness still more heartrending sights in the conflict's final days, when he traveled to the Western Front as an official war artist.

At the behest of the British War Memorials Committee—a government body in charge of commissioning contemporary artworks to create a memorial of the Great War—Tonks set out for France, just days after Bolshevik revolutionaries shot and bayoneted the former Russian tsar Nicholas II, his wife, and his children. Tonks had been tasked with making studies for a painting of a dressing station, where medical personnel first evaluated the wounded and sometimes bandaged or operated on them. Accompanying him on his journey was the American expatriate John Singer Sargent, considered the leading portrait artist of his generation, who would soon create one of the most memorable war paintings of all time.

Tonks, who was a sensitive soul attuned to the suffering of others, may have felt apprehensive about returning to the Front. But there was reason for cautious optimism. Despite gains made earlier that year, the Germans had exhausted themselves during the spring offensive. Through June and early July, they had failed to break through the Allied defense lines in France, due in part to the arrival of American reinforcements. By the time Tonks arrived, German morale had begun to crack. On July 18, French forces in the Marne launched a surprise counterattack, which resulted in a victory for the Allies. It was the beginning of the end of a very long war.

These gains notwithstanding, it had been years since Tonks had witnessed any military action, and the pandemonium of battle rattled his nerves. He wrote to a friend about "a peculiarly vicious gun very near" that fired every three minutes. But he would have to tolerate these distractions if he was to produce any artwork of worth. Both he and Sargent made sketches of a dressing station that treated a steady stream of wounded men, most of them victims of gas attacks. The haunting scene of a parade of bandaged soldiers rendered sightless by chemical weapons—literally the blind leading the blind—eventually inspired Sargent's magnum opus, *Gassed*. From his studies, Tonks produced a similar work showing a different dressing station at the foot of a ruined church. Casualties with bandaged limbs and occasional head wounds crowd the foreground as stretcher-bearers carry more wounded into the scene. As he did at the Queen's Hospital, Tonks made full use of his medical expertise in depicting the frontline management of the injured.

Shortly before returning to Britain, Tonks confessed, "I have seen enough to last my lifetime." He was not alone. By then, the entire world had grown weary of the fighting, even as each side continued to pummel the other. But there was something far more sinister than war on the horizon.

It began with a rasping cough. Annie Elinor Buckler might not have given it a second thought at first—she was too busy tending to patients on the hectic wards of the Queen's Hospital. And, at the age of forty-three, she had suffered many seasonal colds in her lifetime. She had no reason to make the connection to the deadly new virus that was sweeping through Europe that autumn.

In fact, it may have been that very few of the hospital's staff or patients were even aware of the devastation this "Spanish Flu" was causing at home and abroad. When it appeared in the spring of 1918, the military's medical officials were uncertain as to its origins. As the number of casualties grew, they began referring to it as the "three-day fever" to reflect the nature of the virus: three days of incubation, three days of fever, and three days of convalescence. But reports of the virus's spread remained sparse and incomplete, even within the medical community. And there was another problem.

Britain, like most countries fighting in World War I, was subject to media blackouts designed to prevent bad news from affecting public morale. Thus, the first newspaper reports of the virus came from Spain, a neutral country without any wartime media restrictions. There, the outbreak captured public interest because the king himself was one of the early cases. From then on, it became widely known as "Spanish Flu," though it is generally thought that the first cases occurred in a military camp in Kansas after a private named Albert Gitchell reported flulike symptoms on March 4. From there, the virus jumped from soldier to soldier, eventually making its way to Europe, where it began killing en masse.

When the virus first appeared, few people would have imagined that an outbreak of influenza would end up claiming many times more lives than the war itself: between fifty and one hundred million people, civilians and soldiers alike, died over the course of eighteen months. After all, influenza was not new to the British military. Indeed, there had been tens of thousands of cases since the start of the war. But the strain that emerged in 1918—now known as influenza A (H1N1)—was particularly vicious, and living conditions in the army contributed to its rapid spread. Soldiers were bunched together in overcrowded trenches, and after years of fighting, many of them were malnourished and immunocompromised. Formerly healthy young men were transformed into prime targets for the virus. Moreover, there was increased movement of both civilians and troops across Europe due to the war, which allowed the flu to spread more quickly over greater distances.

The pandemic didn't come as a surprise to everyone. As early as 1914, public health officials began voicing concerns that global warfare could introduce new diseases into civilian populations. These experts understood that history's deadliest epidemics occurred when previously isolated populations came into sustained contact with one another. The Plague of Justinian, which struck the Byzantine Empire in the sixth century, traveled with shipments of grain from North Africa, while the Black Death in the mid-fourteenth century made its way from Asia into Europe aboard Genoese trading ships, which docked at the Sicilian port of Messina after a long, dangerous journey across the Black and Mediterranean Seas. The people who gathered on the docks to greet the ships were met with a horrible surprise. Most of the sailors were dead, and those who were still breathing were barely clinging to life.

Unsurprisingly, epidemics have also traveled with armies, such as when the Spaniards introduced smallpox into the "New World" during an invasion of the Aztec and Incan empires. Similarly, the troop movements of the First World War were a perfect vector for disease. Guy Carleton Jones, who would later become Surgeon General of the Canadian Army Medical Corps, warned at the start of the conflict that the "trail of infected armies leaves a sad tale of sickness amongst the women and children and non-combatants. Laws and regulations may govern the conduct of war, but disease and infections recognize no such laws and refuse to signal [sic] out the combatant only." His worst fears were realized four years later when this new and unusually virulent form of influenza began sweeping through both military and civilian populations, claiming millions of lives. "Thus we see that war forces itself on the civilian, on the innocent child, on the non-combatant who stays at home . . . for who can tell, or count up, or even recognise the victims of war when it once places its hand on a country?" Jones asked.

The 1918 influenza pandemic crashed over humanity in three waves. The first began in the spring and might have gone unnoticed had it not been for the second, which arrived in the autumn after the virus mutated into a deadlier variant—just as Annie Elinor Buckler began developing a cough at the Queen's Hospital in Sidcup. Symptoms were so unusual that doctors sometimes mistook the virus for dengue fever, cholera, or typhoid. One witness observed that "[o]ne of the most striking of the complications was hemorrhage from mucous membranes, especially from the nose, stomach, and intestine . . . Bleeding from the ears and petechial hemorrhages in the skin also occurred."

The third and final wave of the pandemic arrived the following spring and lingered until 1920. In its most potent form, the virus could kill as quickly as it could spread. It was said that a person could be well in the morning and dead by evening. Corpses piled up at an alarming rate. In a letter to a colleague, one physician stationed at a U.S. Army camp wrote, "It is only a matter of a few hours [until] death comes . . . It is horrible. One can stand it to see one, two or twenty men die, but to see these poor devils dropping like flies . . . We have been averaging about 100 deaths per day."

There was hardly anyone who wasn't touched by the tragedy. Physicians and nurses struggled to cope as hospitals were overrun with the sick and dying. Many were struck down due to their occupations. Nurse Buckler was among them, but she wasn't the only one of Gillies's staff members to perish during the pandemic. Eleven days after her death, Captain Ernest Guy Robertson—a thirty-three-year-old dental surgeon who had spent two years working at a casualty clearing station in France before being assigned to the Queen's Hospital—also succumbed to the virus.

Unsurprisingly, the most vulnerable were the patients, many of whom were in the midst of lengthy recoveries, their strength already depleted when the virus began invading hospital wards in Britain. Private Abraham Clegg, who had been hit in the mouth with shrapnel, contracted influenza after being sent to the Queen's Hospital for reconstructive surgery. He died, just a few months after enlisting in the army. Reginald Ernest Trease also died during the pandemic after undergoing his nineteenth operation at Sidcup. He was only twenty-nine years old. Others joined the legions of men who had survived the hell of war, only to be felled by this new illness.

Harold Gillies, like so many who lived through the global pandemic, knew countless people who succumbed to the flu during that time, though he himself escaped illness. Only later did he discover that Hippolyte Morestin, the cantankerous surgeon who had locked him out of the operating theater in Paris at the start of the war, was also on the long list of victims. Death struck swiftly and indiscriminately, and those working to save the lives of others were not immune. But, as always, the work at the Queen's Hospital ground on.

Though a pandemic was raging, peace was, nonetheless, at

hand. Following the Allied victory in July at the Second Battle of the Marne, the British, Belgian, French, and American armies mounted a series of pushes to drive back the Germans. This multilateral effort was known collectively as the Hundred Days Offensive. The fighting was heavy and continuous, but at the end of September, the Allies were able to break through the Hindenburg Line—the last vestige of German defenses on the Western Front. The end was now both inevitable and imminent.

⇥ 13 ⇤

ALL THAT GLITTERS

aryl Lindsay was making his way up a ramp toward the dispensary when he crossed paths with Colonel Henry Newland, the surgeon who headed the Australian unit at Sidcup. Newland, a man who often kept his emotions closely guarded, looked as if he were in a trance. As the two men passed one another, Newland mumbled, "Interesting news." It was as if he were commenting on a sports report in the newspaper rather than the end of a major global conflict.

Germany signed the treaty that ended the fighting in the early hours of November 11, 1918. The guns fell silent on the Western Front at 11 a.m., and by the evening, joyous crowds had taken to the streets of London in spontaneous celebration. Newspapers reported on the "surging mass" of people who "wandered aimlessly about, indulging in all sorts of minor horseplay." Men and women, "breathless and hatless," rushed from their homes, their offices, and their shops to cheer. The roads became impassable as the capital sprang to life with the sounds of victory: "[b]ells were rung, trumpets and bugles blown, and there was much banging of tins."

But the jubilant mood was tempered by the grief of the many

thousands who mourned for the dead, some of whom were not yet cold in their graves. Those who yearned for the return of the bodies of their sons, husbands, and brothers would have a long wait. Countless men had been hastily buried in temporary cemeteries near where they fell. Although many soldiers wore standard-issue ID tags around their necks, these often contained only a single metal disc displaying the wearer's name, number, rank, regiment, and religious denomination. When a man died, the disc was removed for administrative purposes, leaving the body with no identification marker. The job of returning the dead to their native countries was, therefore, a difficult one. Exhumations continued with regularity throughout the 1920s, with the French alone shipping an average of *forty* bodies a week to the British during that time. To this day, efforts to locate and identify the fallen of the First World War continue.

The troubles of the dead were over, at least. But those of the vast majority of wounded soldiers were not. Some of them were making their way to Sidcup just as the corridors of the Queen's Hospital were buzzing with news of the Armistice. Outside the dispensary, Newland was imparting his "interesting news" to Lindsay. "[A]s we spoke," Lindsay observed, "he was unconsciously snapping some tubes of tablets in his hands, and the tablets were falling on the ramp." Like so many at Sidcup, Newland was a man "who had seen the war in all its phases, and now it was over." There was an atmosphere of incredulity mixed with relief at the hospital.

Like Newland, Harold Gillies felt unbalanced by the news—as if a light switch had been abruptly turned off. Later, when reflecting on the moment, he could still only muster a few words: "the war ended." But his mind was justifiably elsewhere on the day of the Armistice itself, because he had another, more personal event to celebrate. His wife had given birth to a daughter, Joan, just five days before the war ended. As jubilation broke out on the wards of the Queen's Hospital, Gillies headed to the register office to report the birth of his third child.

As it turned out, 1918 was an unforgettable year of death and suffering, only somewhat tempered by the promise of renewal.

The days may have been growing colder and darker by mid-November, but nothing could cast a shadow over Daisy Kennedy's mood. She was attending a luncheon in Mayfair, which was one of countless events held in celebration of the war's end. Kennedy, an Australian violinist married to the pianist Benno Moiseiwitsch, became absorbed in conversation with a handsome young officer sitting next to her at the table. He had been fighting on the Western Front until the very end and seemed to have escaped the war completely unscathed.

Over the sound of clinking glasses and tinkling silverware, Kennedy mentioned Harold Gillies, a fellow antipodean, whose work was by then renowned the world over. She expressed pride at their shared origins, both being from the southern hemisphere. "You couldn't pay me a greater compliment," the young man remarked between mouthfuls of food.

Kennedy looked up from her plate, startled at the thought that she might have been sitting next to the famous surgeon all along without realizing it. "But you are not Major Gillies?" she asked.

The officer, with his flawless face, responded, "No, I was one of his patients."

"I was so moved that I couldn't speak," Kennedy later recalled. "His face bore no sign of ever having been under a surgeon's hand."

>-=<

On June 28, 1919, the afternoon sun was streaming through the tall, arched windows of the Hall of Mirrors at the Palace of Versailles. This was the venue for the signing of the treaty that would formally conclude the First World War. A handful of disfigured French soldiers, dubbed the *Délégation des Mutilés*, made its way into the glittering jewelry box of a room. They had been invited by their prime minister to stand as a visual testament to the grisly nature of the conflict. Leading them was Albert Jugon.

Five years earlier, Jugon had been left for dead on the edge of a trench during the first weeks of the war. A shell splinter had torn away half his face and throat, smashed his jawbone, and punctured his right eye. So dire was his condition that a priest had given the young man the last rites on the battlefield.

Until February of that year, Jugon had been at Paris's Val-de-Grâce military hospital under the care of Hippolyte Morestin. Before he had succumbed to influenza, Morestin had chosen Jugon from among his hundreds of patients to attend the ceremony. In turn, Jugon selected the other four men who would accompany him on the journey to Versailles: Eugène Hébert, Henri Agogué, Pierre Richard, and André Cavalier.

As these men shuffled into the gilded gallery, they joined hundreds of dignitaries, journalists, and members of the public, all of whom had come to witness the signing of the peace treaty. Jugon and his comrades were not the only reminders of the devastation wrought by the war. Despite the joyous nature of the event, most attendees wore black, out of respect for the millions of people who had died over the course of the conflict. The Délégation des Mutilés took up a position behind a small table near the center of the hall. Their location meant that the plenipotentiaries had to pass in front of them in order to sign the historic document. Ironically, these soldiers, who had spent much of their recovery evading their reflections, were no longer able to avoid the sight of their own disfigured faces—surrounded as they were by 357 mirrors. Nor could anyone else fail to notice the men's damaged features.

Jugon and his companions took their seats while people flitted about the gallery, looking for delegates to sign their commemorative copies of the treaty. "[I]t was amusing to notice the eagerness with which famous men wandered from one end of the very long room to the other in search of the autographs of equally famous men," one journalist wrote. Autograph hunters mingled with photographers trying to capture every moment of the afternoon. "Nearly everyone seemed to have a camera, and there was a perpetual taking of snapshots in every part of the hall," another reporter observed.

The leaders of the top four Allied nations made their way to the palace through throngs of onlookers lining the streets. They included Georges Clemenceau, the French prime minister; Woodrow Wilson, president of the United States; David Lloyd George, prime minister of Great Britain; and Vittorio Orlando, prime minister of Italy. Known as the "Big Four," these men were the leading architects of the peace treaty. Other delegates from these same countries—as well as emissaries from nations affiliated with the Allies—played peripheral roles, while representatives from the Central Powers had little say in the framing of the accord. In time, its harsh terms would lay the foundations for a second and even more devastating global conflict.

As the leaders arrived, they took their seats alongside other delegates from the Allied nations. Shortly after three o'clock, an uneasy silence fell over the room. Outside, the rumble of a motor sounded in the distance. Inside, attendees looked anxiously at one another, whispering: "Here they come."

The German representatives—Foreign Minister Hermann Müller and Minister of Colonial Affairs Johannes Bell—entered the Hall of Mirrors flanked by French, British, and American officers. Once the two men had been seated, Clemenceau stood up to make a speech, at the end of which he invited the Germans to sign the treaty. The two men hurried forward, only to be stopped in their tracks by the official interpreter, who began translating the prime minister's speech into English. When the interpreter reached the words "German state," a voice protested, "German Reich," and the interpreter promptly repeated the correction.

The observance of ceremony notwithstanding, it was perhaps inevitable that any official act marking the end of such a protracted cataclysm would seem anticlimactic. One journalist wrote that the "ceremony was curiously unimpressive" despite the magnificent surroundings. After the Germans had signed the document, the other delegates lined up to do the same. At last, Clemenceau and the French delegation rose to sign the peace treaty. As the prime minister made his way to the center of the room, he paused in front of the Délégation des Mutilés. "You have suffered a lot," he told the five men whose lives would never be the same. Clemenceau then gestured to the table where the historic document was waiting to be signed, and added, "[B]ut here is your reward." With the ink of the signatures drying on the page, the peace treaty was complete.

It took just thirty-seven minutes to bring an official end to four years of global war.

>-■-<

Far from the grandeur of Versailles, Harold Gillies laid down his sketches and stubbed out a cigarette. He stood up from his desk, giving his case notes one final glance. As he made his way across the grounds of the Queen's Hospital toward the makeshift huts that housed his patients, Gillies visualized the stages of the forthcoming procedure in his mind. Although he had operated on hundreds of faces, Gillies knew that complacency was the enemy. Plastic surgery might be guided by a general set of principles, but it was—above all else—a highly specialized art form.

He opened the door to the operating room and made his way to the washbasin, where he began vigorously scrubbing his hands and forearms. Next, he donned a pair of thick rubber gloves. He then turned to the patient lying at the center of the bright and airy room. The young man had so nearly made it to the end of the war unscathed.

"Don't worry, sonny," he said, offering a gentle smile to the war-ravaged soldier on the table. "You'll be all right and have as good as face as most of us before we're finished with you."

EPILOGUE:
CUTTING A PATH

The Treaty of Versailles had ended the war, but for many dis-figured soldiers, a long road of painful surgeries lay ahead. Harold Gillies's service with the Royal Army Medical Corps officially concluded on October 8, 1919, though he contin-ued operating at Sidcup for another six years. It would be the first postwar chapter in a career that would yield still more challenges and breakthroughs.

The number of people at the Queen's Hospital began to shrink as staff returned to their home countries, taking with them their patients and their medical records. Apart from a handful of sur-geons, many returned to their old jobs, giving up plastic surgery altogether. Henry Newland, who headed the Australian unit, re-turned to Melbourne, where he resumed the practice of general surgery. Others followed similar paths. In the spring of 1920, the Ministry of Pensions took control of the hospital and began admit-ting general medical and surgical cases, even as Gillies continued to rebuild the faces of the remaining soldiers in his care.

Despite the dwindling number of plastic cases, the Queen's Hospital remained a hub of creativity and innovation. It was during this time that the anesthetist Ivan Magill pioneered endotracheal anesthesia. He also designed and adapted a host of instruments, including angled forceps that could be used to guide a tracheal tube into the larynx. The "Magill forceps" are still used in operating rooms today.

In 1925, the eight remaining facial patients at Sidcup were moved to Queen Mary's Hospital, approximately twenty miles away in Roehampton. To Gillies, it felt like the end of an era. One day, shortly before Gillies left Sidcup for good, Thomas Kilner—a surgeon who had trained with him during the war and who would become an important plastic surgeon in his own right—found his mentor in his office in Frognal House. "With tears in his eyes, [Gillies] expressed the fear that all that had been gained by those years at Sidcup would be lost unless some of us continued to specialise in the kind of surgery we had been doing," Kilner later recalled.

The future of plastic surgery was uncertain.

After the war, Auguste Charles Valadier donated all his records—including wax and plaster casts, molds, photographic negatives, and prints—to the Royal College of Surgeons in London. Although he operated on a smaller scale than Gillies, Valadier was an important pioneer of early plastic surgery, and his records attest to the incredible reconstructive work he undertook in France during that time.

Once the dust had settled, Valadier returned to Paris, where he opened a new dental practice. Unfortunately, he also became addicted to gambling. In the late 1920s, he retired from active practice but continued to run up debts. After developing a blood disease (possibly leukemia), he died penniless on August 31, 1931, at his villa in Le Touquet, on the coast of Normandy.

His widow, Alice, was left to fend off his creditors. When the

French authorities threatened to repossess her home, she traveled to the British Embassy in Paris to seek reimbursement for the work Valadier had done free of charge during the war. After a good deal of resistance, the Officers' Association eventually granted her forty pounds. This meager amount hardly made a dent in Valadier's considerable debts. It was only thanks to a generous gift from one of his former patients, an Indian maharajah, that Alice avoided destitution.

Valadier was one of only two dentists knighted for his service during the First World War. Despite this honor, his contribution to plastic surgery went largely unacknowledged in his own lifetime.

While at Aldershot and Sidcup, Henry Tonks created seventy-two pastels of soldiers that captured their faces before, during, and after reconstructive surgery. In Germany, drawings and photos of disfigured soldiers were published as anti-war propaganda. In Britain, however, Tonks's portraits were never widely disseminated to the public. The disfigured face remained largely absent from British wartime art during the twentieth century, apart from the unusual portraits created at the Queen's Hospital.

Before the war had ended, Tonks was appointed Professor of Fine Art at the Slade School—a post he held for twelve years until his retirement. He thrived in his role as an instructor and seemed oblivious to the fact that his students lived in fear of his critical eye and sharp tongue as he snaked his way around the classroom, crushing their confidence with a few sour words. Tonks's biographer, Joseph Maunsell Hone, observed that "[i]t horrified Tonks to find he had it in his power to inflict so exquisite a torture on a human being." But plenty of his pupils, Daryl Lindsay included, would later testify that it was Tonks's scathing critiques that made them the artists they eventually became. When he finally retired in 1930, Tonks was so moved by the farewell ceremony that he felt that he would break down should he ever pass through the gates of the Slade School again. "I have loved my students," he confessed in a letter to a friend.

After retiring, Tonks was offered a knighthood. He declined on account of having no interest in fame or fortune, and he didn't feel that he needed recognition for the role he had played during the war. Art was all that mattered to him. "[M]y painting is more than my amusement, it is my life," Tonks declared shortly before his death in 1937.

As it turned out, Percy Clare never did make it back to the front after he was discharged prematurely from Sidcup on that cold winter's night in January 1918. Nor was he returned to the cozy, warm wards of the Queen's Hospital. After he left Sidcup, he headed back to Dover to report for duty. It wasn't long afterward that his jaw locked up—a condition brought on, no doubt, by the fact that the reconstructive work had been cut short. Despite Gillies's best efforts, not all the men who passed through his wards enjoyed the happiest of endings.

Clare was eventually sent to another hospital, where he underwent further work on his jaw. He remarked that "it was not like Sidcup but quite a haven of rest after [being in] the barracks." He continued to speak highly of Gillies and the Queen's Hospital in his diary, even as he was being treated elsewhere. On July 10, 1918— four months before the war ended—Clare wrote, "I handed in my khaki suit and all that would identify me as a Tommy and went home in my own civilian clothes a free man." He had been honorably discharged due to the severity of his medical condition.

After the war, Clare received the British War Medal and Victory Medal for his service. It's unclear whether he underwent further operations on his jaw later in life. His diary ends where the war ends. Clare died on April 30, 1950, at the age of sixty-nine, leaving behind his son, Ernest, and his wife, Beatrice.

*The transformative work carried out by Harold Gillies and his col-*leagues at Sidcup was not recognized in the years immediately after the war—a snub that did not go unnoticed by Sir William Arbuthnot Lane, who had been instrumental in helping Gillies establish the Queen's Hospital. "[T]o my amazement, such monetary and titular awards were allotted only to . . . the fighting generals," Lane complained. "Men [who] save life never get the same appreciation and reward as those whose business it is to destroy it."

This oversight was eventually rectified in June 1930, when Gillies was knighted for his service during World War I. Gillies's fourth and last child, Mick, would eventually follow his father into medicine, spending most of his career in the tropics studying the transmission of malaria. But he once recounted his boyhood memory of being summoned to his headmaster's office as the news of his father's knighthood broke. "You won't be writing to Major Gillies anymore," the headmaster explained, while waving a copy of *The Times* in front of the boy's nose. "You'll be sending your letters to Sir Harold Gillies in the future." Those closest to Gillies felt that he had been overlooked for too long. The essayist E. V. Lucas congratulated his friend in a letter, writing: "Dear Facemaker, — I am so glad that the King has come to his senses." Lane could hardly conceal his annoyance with the delay: "Better late than never," he wrote.

Despite its tardiness, Gillies welcomed the news, though he reflected in an interview that he regarded his knighthood not as "a personal honour but as one shared by all those who had been with me in the pioneer work." His patients, on the other hand, saw it as an individual victory for the man who had spent the entire war performing daily miracles.

Shortly after the public announcement, letters began arriving by the dozen. "I can never forget your wonderful kindness to me and all that you have done to make my life worth living," one man wrote to Gillies after the doctor was knighted. "I am looking so well that people are beginning *not* to believe it when I tell them that I was nearly burnt to death eleven years ago." Another patient wrote that people still didn't believe that part of his upper jaw was ever missing. Many wondered what their lives would have become had they not found their way into Gillies's skilled hands. As another correspondent said, "When I think of myself before I came to you, my gratitude knows no bounds." Gillies may have restored these men's faces, but figuratively, at least, they remained faceless due to their great number. One soldier remarked, "I don't suppose for one moment that you remember me, for I was only one of many, but that matters little, for *we* remember you."

>–•–<

As Gillies's postwar caseload at the Queen's Hospital shrank, he looked to expand his practice in the civilian world in order to make ends meet. But while the war may have been over, the fight to have plastic surgery recognized as a legitimate branch of medicine had only just begun. Gillies recognized the risks of going into private practice. "To venture into this rather new field of civilian plastic surgery was certainly a gamble," he later confessed.

It wasn't as if Gillies didn't have options. Sir Milsom Rees, his former employer who had made his fortune spraying the throats of famous opera singers, had never filled Gillies's position at 18 Upper Wimpole Street. Gillies had an open invitation to return to his old practice. "It meant reassociation [*sic*] with royalty and certain financial success," Gillies wrote. Nevertheless, he resisted the temptation to retreat to what was comfortable and easy. His decision raised more than a few eyebrows within the medical community. "Don't be a fool," Rees scoffed. "You'll spend your whole life dealing with deformities"—as if this would be a waste of surgical talent.

The world's leading medical journal, *The Lancet*, took an even dimmer view of Gillies's prospects. A piece appeared in which the author argued that "the time may yet hardly be ripe for a plastic surgery department at a general hospital." But Gillies was never one to shrink from a challenge. He was determined to prove the establishment wrong by making a successful career out of being a plastic surgeon. And so he took a consulting room at 7 Portland Place in London. "Name plate up. Secretary installed," he wrote in a letter to Tonks. "Now all I want is a few patients willing to place themselves in the hands of a surgeon crazy enough to nail his fortune—and that of his wife and four children—to the mast of plastic surgery."

*In order to appeal to a broader clientele, Gillies had expanded his prac*tice to include cosmetic surgery even while he continued his work at Sidcup. This was done partly out of necessity in order to attract more paying customers. But there is no doubt that Gillies was also intrigued by the prospect of new surgical challenges. "Reconstructive surgery is an attempt to return to normal," he observed, while "cosmetic surgery is an attempt to surpass the normal." Gillies was not the first to perform what he sometimes termed "beauty surgery." Indeed, there had been a growing interest in cosmetic surgery ever since the discovery of anesthetics and development of antiseptics in the latter half of the nineteenth century, both of which made elective procedures safer and less painful.

The work of Jacques Joseph—the German Jewish surgeon who got his start performing rhinoplasties for Jewish clients before the war—mirrored that of John Orlando Roe, a surgeon from New York who, in 1887, had developed an intranasal technique to alter the tip of the nose without creating external scars. A handful of other surgeons scattered around the world had also taken a keen interest in the burgeoning field of cosmetic surgery at around this time. Chief among them was Charles Conrad Miller, who is often credited as one of the first cosmetic surgeons fully dedicated to "beautification" in the United States. He began performing procedures on the face at the turn of the twentieth century, with mixed results. In his book, *The Correction of Featural Imperfections* (1907), he details surgical methods for creating dimples, plumping lips, removing crow's feet, and pinning back protruding ears. Although many within the medical community considered Miller a charlatan, at least one medical journal characterized him as someone who had built his reputation "upon honest effort in the uplifting of practice of this kind."

Due to his work during the war, Gillies brought a wealth of experience to his cosmetic practice that surgeons like Miller couldn't. In fact, he felt strongly that no practitioner could consider himself a plastic surgeon unless he had mastered both reconstructive and cosmetic surgery. "It is easier to reduce than produce [as with reconstructive surgery], but in [cosmetic] surgery it is nearly always necessary to remould after reduction," he warned. "Thus anyone can cut off a bit of nose or breast, but not so many can turn out a satisfying result."

Initially, business was slow to take off. Even when Gillies treated private clients, he discovered that it was not always easy to collect payment from them. On returning from a trip to America in 1919, Gillies met a charming woman aboard ship who expressed unhappiness with her long, beaky nose and wished to undergo rhinoplasty to reshape it. The operation was to be funded by her lover, who was a prominent financier in London. In secret, however, the woman confided to Gillies that she was in love with another man, named Hugo. At separate times, both Hugo and the financier met with Gillies to discuss the future shape of her nose. "Hugo was for the Grecian type—the moneyed one was for a rather turned-up [nose]," Gillies wrote. With one eye to business, Gillies ended up giving his patient a retroussé nose, with the tip turned ever so slightly upward. Unfortunately, when the time came for the bill to be paid, the financier refused on account of the fact that the woman had "set out to conquer new fields, leaving him and poor Hugo in the lurch"—not to mention Gillies.

Failure by his clients to pay their bills was the least of Gillies's problems during the early days. Once, he was consulted by a woman who had allowed a beautician to inject paraffin wax into her face. The wax had begun to shift underneath her skin, and she was experiencing painful ulcerations all over her face. Although he had no experience in removing paraffin wax, Gillies agreed to help. Unfortunately, his first attempt to correct the issue was unsuccessful, and Gillies found himself one afternoon face-to-face with a very irate husband. "These things take time, you know," he offered as a weak explanation. The man, crazed with anger, whipped out a revolver from his pocket and pointed it squarely at the surgeon, whom he blamed for "ruining" his wife's face. Recalling the harrowing incident, Gilles quipped, "I have since been informed by my more experienced colleagues that a well-tailored bullet proof vest can be worn with comfort."

Gillies was always able to find the humor in his work. He often regaled people with anecdotes from the early days of establishing his private practice. When a young woman who had been disfigured in a car accident turned to him for help, he agreed to rebuild her face. Gillies told the woman's husband that the work would require a large skin graft and suggested that it be taken from his buttocks. The man readily consented. Years later, Gillies bumped into the same man, who thanked him profusely for restoring his wife's appearance. The husband added that he didn't regret donating skin from that part of his anatomy. Indeed, quite the contrary. "For whenever my mother-in-law spends the weekend with us and kisses my wife good bye [*sic*]," he cheerfully proclaimed, "I always feel I'm getting my own back."

It was clear that Gillies enjoyed performing cosmetic procedures. Although he was careful to wait until he was approached for advice, he did on rare occasions give a potential patient a nudge. "I do plead guilty to casting one fly for a patient," he confessed. While on a fishing trip, he met the daughter of the innkeeper, whom Gillies described as a "comely lass with a fearsome nose." As it was her job to dust the sitting room when he was out fishing, he decided to leave a book containing before and after photos of noses he had reconstructed. "The trout rose was hooked and returned to the water with an undersized nose," he wrote with delight.

As business took off, Gillies found himself at the pointed end of criticism from those who saw his cosmetic practice as nothing more than a money-making venture. Frances Steggall, a nurse who worked with Gillies during the war, remembered a colleague remarking to her that "Sir Harold's face-lifting operations on la-dies . . . would soon make [him] a fortune." Steggall bristled at the idea that Gillies was only motivated by money. She told the other nurse a story about a young man in the East End of London who could not secure the employment he was seeking due to the exten-sive burns on his face. "Sir Harold undid his scars," Steggall told her, "[he] made him presentable and [the young man] got the job he wanted." The man never received a bill for the work Gillies did.

He wasn't the only patient to receive free care from Gillies after the war. During a golf tournament in Sandwich—where Gillies had once jumped off the train to try out for Cambridge Univer-sity's golf team—he was approached by a local doctor who had pushed through the crowds to ask him to examine one of his pa-tients. "As there seemed to be some time before we could tee off again," Gillies recalled, "I went with him to meet Ernie, a fifteen-year-old caddie with a very tight upper lip." When he greeted Er-nie, the boy hung his head in embarrassment, trying to conceal his disfigurement. "Without looking into his mouth I could imagine the short, scarred palate trying in vain to reach his pharynx." Gillies decided then and there to help Ernie. After successfully repairing the boy's mouth, Gillies could not be found on the golf course at Sandwich without the smiling caddie at his side.

Gillies could be overly generous. Those around him noted that he "would give a young assistant £50 where other consultants would have thought £5 adequate." Gillies was also quick to help friends and family members who were strapped for cash. Most of all, he loathed discussing money with his patients. He left this unpleasant task to his private secretary, "Big Bob" Seymour, who had come to him at Aldershot after his nose was shot off at the Battle of the Somme. "Talk it over with my secretary," he would say to his patients as he waved them out the door. "Mr. Seymour has my standard fees."

Big Bob did his best to protect Gillies's financial interests. But his boss, who had so many other talents, proved that making money was not one of them. Gillies would offer fee reductions to all sorts of patients, such as the "woman who has children to feed and needs a youthful look or improvement to keep her job," or the wife who feared she might lose her husband due her aging appearance. "She should be warned that merely lifting her face, breasts, or nose, or all three, will not hold her husband," he wrote. "Every effort, however, should be made to stack the cards as far as possible in her favour." This involved reducing his fee—sometimes considerably—in order to make the "required" work more affordable to the patient. Decades after the war had ended, Gillies corrected a former colleague's assumption about the wealth he had accumulated in his private practice, writing, "I am afraid your ideas of the 'fortune' I have made are erroneous. I don't think I am worth any more today than I was in August 1914."

Whether or not there was money to be made in plastic surgery, Gillies was aware of the questions from the public as well as those inside the medical community concerning the justifiability of cosmetic procedures. He claimed to have no issues about operating for vanity's sake alone. "[I]f you are not going to be vain about the nose I'm going to give you then I have no interest in doing it," he told patients. Nevertheless, it was clear that Gillies felt conflicted by his work at times. "Often while lifting a face I have had a feeling of guilt that I am merely making money, and yet to see the lasting pleasure that often follows makes me wonder who we are to refuse a patient," he wrote. More than most people at that time, Gillies understood that "deviations" that might seem trivial to a casual observer could often be a source of distress to the afflicted. He wondered if cosmetic surgery was therefore justified by the "little extra happiness [it brings] to a soul who well needs it." In the end, he concluded that it was.

While many clients looked to Gillies for simple enhancements, others sought him out for very different reasons. After X-rays were discovered in 1895, one of their more faddish applications was the removal of unsightly hair. This continued into the 1950s, and a study of 368 patients in New York in 1970 found that more than 35 percent of radiation-induced cancers in women could be traced to X-ray hair-removal.

Gillies was consulted by a woman who had undergone this treatment and had developed ulcers and carcinomas all over her face as a result. A surgeon had removed her lower jaw and bottom lip, which left her tongue hanging down her neck. The surgeon later told Gillies, "While I'm certain that I removed the cancer, I felt that every nurse in the theatre would have gladly put a scalpel between my ribs for giving this woman such a mutilation." The woman was traumatized by her ordeal, and shortly after Gillies began working on her face, he found her hanging out of a hospital window with two terrified nurses clutching at her. Eventually, Gillies was able to repair the damage using tubed pedicles and grafts to reconstruct the jaw. Little by little, her depression lifted as her facial features were restored.

There were even more arduous surgeries during those postwar years, such as the operations performed on Mrs. Brown, who had fallen face-first into an open fireplace while suffering an epileptic seizure. She had been holding her four-month-old daughter at the time of the accident and lay there, unconscious, for hours, slowly roasting. Both she and her child were gravely burned as a result, and her daughter's leg was bent so far back that the heel almost touched her bottom. The pair were rushed to a hospital in Ayr, on the southeast coast of Scotland. Their conditions were so critical that doctors questioned whether the two would even survive.

Miraculously, however, they slowly recovered. Writing about the accident years later, Mrs. Brown's daughter remembered that "most of the flesh of [her mother's] face had dropped off." Her lips were completely gone, and her gums had retracted under the scorching heat, which made her teeth look long and jagged.

Mrs. Brown was eventually sent to London for treatment. There she met Harold Gillies, who, despite being a seasoned veteran, was taken aback by what he saw. "Burns received in war and occupational accidents may be overwhelming in number but none come in worse shape than [hers]," he wrote. He marveled that anyone could survive such an ordeal: "When this poor facial skeleton was brought into my clinic . . . the back wall of her frontal sinuses and the scarred, opaque round objects once her eyes stared me in the face."

The hardest decision was not what he should address first, but whether he should do anything at all. Gillies realized that there was a possibility that he might put this woman through more pain for little or no gain. Even worse, he could cause further damage. "What a harrowing decision confronted me!" Gillies thought as he contemplated the case. "She sat there quietly while I studied her, weighing in my mind if it were justifiable to accept the challenge to try to make her a face." The scale of the task before him was almost unimaginable. Not only had Mrs. Brown lost all the skin on her face, but the lining and musculature had also been destroyed in the accident. "But where to begin?" Gillies asked himself. "The problem was staggering, and the only possibility was to make a start by doing something positive."

Gillies knew that he needed to graft skin onto the raw surfaces of Mrs. Brown's face before he could begin rebuilding it. Unfortunately, the usual donor sites of the neck and forehead had already been damaged. He therefore decided to import skin via a tubed pedicle from her abdomen, which he first attached to her wrist and later to her neck. He then used this skin to create eyelids, cheeks, and a nose—the last of which also received a cartilage transplant. Next, he used local flaps to create a lining for the eye socket and mouth. He also created eyebrows using a visor flap from her scalp "to break the monotony of her flat pedicled face."

The biggest challenge for Gillies, however, was the mouth. "In all tube pedicle mouths the absence of musculature and elasticity renders them unstretchable," he explained. Gillies had two choices. He could either make the mouth small and presentable, which would prevent the fixture of dentures, or, he could create a "wide [mouth] of jack-o'-lantern dimensions" that would allow the patient to have teeth. Gillies chose the second option. In total, Mrs. Brown underwent approximately thirty operations. Before she left the hospital for home, Gillies arranged for her to visit the Elizabeth Arden salon, where she was given cosmetic products to boost her morale. At the end of it all, her daughter wrote, "She was never again pleasant to look at, but she did have a face of a kind."

Gillies was proud of the work he did on Mrs. Brown, not just because he was able to give her back a sense of identity, but because he hoped that his success would serve as an inspiration to other doctors. "Throughout her entire reconstruction there was always the feeling in the back of my mind that in spite of the inevitable shortcomings, the fact that the repair had been attempted would serve as encouragement to any future surgeon faced with a similar catastrophe."

Although he had been able to restore some semblance of a face to Mrs. Brown, Gillies could not prevent her untimely death a few years later from another epileptic seizure. He was playing golf when he received the news. "Poor Mrs. Brown," he muttered to himself as followed his ball down the fairway.

Mrs. Brown's daughter had fared better. As she grew from a baby into a toddler, her twisted leg straightened itself. "All that can be seen now is the scar—thick and unlovely but partly hidden by even a short skirt and causing me only a little more cramp than perhaps is usual," she wrote later in life. She wanted her mother's story told because she felt it reflected well on the transformative power of plastic surgery. "[C]onsidering that Sir Harold built this face from her body on to almost bare bones, surely <u>this</u> case was one of his greatest?" she wondered.

Gillies's reputation began to spread far beyond the shores of Britain. Students from around the globe visited him, hoping to learn from the master. During operations, he would identify his assistants by where they hailed from rather than by their names. "If you were starting this case, how would you set about it, Mexico City?" he would ask, putting similar questions to "Johannesburg," "Oslo," "Newcastle," and "Rio." Gillies was accommodating to a fault—a trait he blamed on his wartime experiences in France. "The disappointment when Morestin closed the door on me has probably been responsible, in a way, for my leaning over backwards to discuss plastic problems with students," he reflected.

But it was more than just a collaborative bent that led Gillies to throw open his doors to aspiring plastic surgeons from around the world. He enjoyed being surrounded by "young eager minds that have not yet learned not to hope and are oblivious to the limitations." It was their enthusiasm and, at times, their naïveté that drove him to push the boundaries of what could be accomplished in the operating room, as he had done during the war when he, too, was youthfully oblivious to the limitations of his chosen field.

Gillies's quest to legitimize plastic surgery did not end with the molding of eager medical students into plastic surgeons. Two years after the war, he published his first major work on the subject, *Plastic Surgery of the Face*, in which he described the key skills and techniques he had honed and the brutal lessons he had learned during the war. This was not the first book on plastic surgery ever written, but the sheer volume and variety of cases it presented made it one of the most valuable at the time of its publication.

>-■-◄

Shortly before one o'clock in the morning on a starless May night, a high-explosive bomb hurtled through the London sky. This was by no means unusual in the dark days of 1941, when the German aerial campaign that had pounded the capital for eight straight months was just coming to an end. But this particular shell tore into the Royal College of Surgeons in Lincoln's Inn Fields. As the dust settled and the fires were extinguished, it became apparent that a priceless collection of historical artifacts had been destroyed, including over six thousand anatomical specimens.

Also lost in the air raid were many of Harold Gillies's case notes, which had been stored in the building since the end of the First World War. Even when distilled into surgical records, it was as if Gillies's patients could not escape the shock waves of war. And yet an important part of their surgical legacy had survived. Peering out from among the rubble were some of their haunting portraits, drawn by the sensitive hand of Henry Tonks.

Gillies's talents and hard-won skills were called upon during the Second World War, when he was fifty-seven years old. But his efforts were largely eclipsed by those of his cousin, Archibald McIndoe, whose reconstructive work on the burned Royal Air Force pilots of the "Guinea Pig Club" brought him international attention. Gillies had introduced McIndoe to the "strange new art" of plastic surgery himself, and it was on his recommendation that McIndoe became a civilian consultant to the Royal Air Force in 1938. McIndoe would improve upon techniques that Gillies had invented during World War I, while developing some of his own for the treatment of badly burned faces.

The attention that McIndoe's burned pilots attracted rankled some of Gillies's former patients. In a letter, Horace Sewell wrote, "[Y]ou know it made my blood boil when I used to read during and after the last war people calling themselves the guinea pigs. The real ones were to be found in Sidcup over 20 years before, and I will go as far as to say that there were far worse cases of burns by liquid fire in those days."

Just as the war was ending in 1945, Michael Dillon approached Gillies with an unusual request. Dillon, who had been assigned female at birth, had been unhappy with his gender since childhood. When he was seven, a family friend joked that she would take him to the blacksmith to be made into a boy. "I had taken her seriously in my delight and excitement, only to be reduced to tears when I found that such a thing could not be after all," he wrote. As Dillon grew older, these feelings only deepened. In the late 1930s, he began taking testosterone pills and had his breast tissue surgically removed. But Dillon wished to complete his transition, and in order to do that, he needed the help of a truly innovative surgeon.

In addition to rebuilding faces, Gillies had also been conducting genital reconstruction on injured soldiers during and after World

War II. Because of this, he was better placed than most to take on Dillon's complex case. Although a handful of surgeons had performed successful vaginoplasty procedures on trans women, no one had yet accomplished its female-to-male equivalent. Indeed, many surgeons would have deemed it impossible. Some may have even believed it unethical. Nevertheless, the legal system was in Dillon's favor. While there were laws that prohibited the removal of a penis, there was no legislation that prevented an individual from adding one. Gillies—who never shied away from a surgical challenge—agreed to carry out a phalloplasty, or the construction of a penis, on Dillon. His decision came as welcome news. "The world began to seem worth living in after all," Dillon later wrote.

Gillies gave his new patient a false diagnosis of acute hypospadias, a birth defect that results in the misplacement of the urethral opening. This was done to protect Dillon's identity as a trans man on his visits to the clinic. Over a span of several years, Gillies operated on Dillon thirteen times. By rolling a tube of tissue on the abdominal wall to produce a urethra, and then surrounding this with another tubed pedicle, Gillies was able to construct a penis. In 1949, Gillies became the first surgeon to complete a successful phalloplasty on a trans man. His pioneering technique laid the foundation on which modern phalloplasty is based.

Dillon was delighted with the results. "How different was life now!" he effused. "I could walk past anyone and not fear to hear any comments for no one looked at me twice." In the years following the surgery, the two men became friends. Dillon visited Gillies at his clinic whenever he was in the area. "He always seemed glad to see me," Dillon wrote, "and invariably reiterated that he was delighted he had undertaken my surgery since it had been so worthwhile. There were many who would not have and my debt to him can never be repaid."

In 1958, British journalists outed Michael Dillon against his wishes. A media frenzy ensued, and he fled to India, where he eventually became a Buddhist monk. In the midst of this turmoil, Gillies wrote to his former patient with words of comfort and encouragement. Dillon later reflected that the plastic surgeon's "one aim had always been to make life tolerable for those who either Nature or man had ill-treated without regard to conventional views and to many a one he must have given renewed hope and a new start." Some people may not have been able to accept Dillon as a man, but Harold Gillies was not one of them.

*Long after the guns had fallen silent on the Western Front, the battle-*field surgeon Fred Albee wrote, "no evil ever befalls the race without bringing with it some good." Among these goods were the medical advances spurred by the carnage of the war. These advances, while offering second chances to individuals, were equally important to humanity as a whole.

Harold Gillies never tired of pushing the limits of what surgery could accomplish. He knew that even his most radical innovations would eventually be surpassed: "one day surgeons will do something further in the way of making new bits and pieces [for patients]." In 1944, Gillies put forward the idea for a professional body that would direct the development, uphold the standards, and safeguard the interests of this burgeoning specialty. Two years later, he became the first elected president of the British Association of Plastic Surgeons.

Decades after he wrote *Plastic Surgery of the Face*, he set out to write a second, more comprehensive book on the subject. This time he enlisted the help of his American protégé D. Ralph Millard, Jr., whom he had met by chance on one of his many trips to the United States. On arriving in Britain, Millard was greeted by Gillies's colleagues, who teased, "We have been placing high odds against your being able to pin the old man down to complete a book or even a total chapter." They were not wrong about the plastic surgeon's talent for procrastination. Millard soon discovered that Gillies was as faithful in life as he was in surgery to his first and everlasting principle: "never do today what you can put off until tomorrow." Despite Gillies's attempts to distract Millard from the task at hand, the duo eventually published *The Principles and Art of Plastic Surgery* in 1957. Shortly afterward, Gillies presented both volumes, specially bound and inscribed, to the queen mother, who told him that she was "proud to have received them from you as pioneer of this great and healing branch of surgery." The book remains, even to this day, one of the seminal works on the subject.

Gillies's transformative work during the war marked a turning point in medical history, as he opened the door for a new generation of plastic surgeons concerned not just with function, but also with aesthetics. Some plastic procedures, such as rhinoplasty, predated the war, but it was under Gillies's direction that old techniques evolved, and new ones were imagined, tested, and standardized. His importance to the development of plastic surgery in general, and to his patients in particular, is difficult to overstate. "As a result of [the] efforts of one man," the surgeon Neal Owens wrote shortly before Gillies died, "the world has become a better place in which to live and existence in the present troubled world has for many been made more worthwhile."

Harold Gillies was a genuine visionary in his field, and his drive to do his best for those in his care stayed with him until the very end. On August 3, 1960, he suffered a minor stroke while operating on an eighteen-year-old girl whose leg had been shattered in a car accident. He died a month later, at the age of seventy-eight.

>-<

As Gillies predicted, plastic surgery has evolved in ways that even he could not have imagined when he began advocating for its recognition shortly after the end of World War I. To alter their physical appearance today, people can choose from a seemingly infinite number of cosmetic procedures: breast implants, tummy tucks, liposuction, face-lifts, and more. The public's growing fascination with plastic surgery—partly driven by the proliferation of reality television programs featuring plastic surgeons and their patients—has created a boom in an industry that is now worth billions of dollars.

While interest in cosmetic procedures is at an all-time high, reconstructive surgery aimed at repairing and restoring function to those affected by congenital abnormalities, trauma, or disease continues to be a mainstay of the discipline. One of the more recent developments is the "face transplant," which involves replacing all or part of a person's face using tissue from a donor. The procedure, which is considered "life-enhancing" rather than "life-saving," continues to frame the problem of facial difference as an individual deficit rather than a societal one. Its necessity is driven at least partly by prejudicial attitudes toward disfigurement that even Corporal X, who broke off his engagement to Molly after catching a glimpse of his reflection in a mirror, might recognize today. Whatever the driving force, however, face transplants have undoubtedly added value to some patients' lives, allowing them to eat solid food, to breathe independently, and even to smell for the first time in years.

In 2005, surgeons performed the first partial face transplant in Amiens, France—where the Hundred Days Offensive began in 1918—on Isabelle Dinoire, whose dog had bitten off part of her nose, chin, and lips. Five years later, surgeons in Spain performed a full-face transplant on a man who was injured during a shooting accident. Similar operations followed in various countries around the world. These first successes engendered great public curiosity and even greater technical leaps forward in surgery.

In 2017, an army of masked surgeons crowded around a small instrument table in an operating room at the Cleveland Clinic in

Ohio—a hospital founded in part by three doctors who had served together during World War I. There was a mixture of exhaustion and awe in their eyes as they stared down at what looked eerily like a rubber mask.

Hours earlier, they had begun meticulously removing the face of a woman who had died from a drug overdose in order to transplant it onto twenty-one-year-old Katie Stubblefield, who had suffered severe facial trauma due to a self-inflicted gunshot wound. Stubblefield was about to become the youngest patient to undergo such a revolutionary operation. Although this was the third face transplant undertaken at this hospital, it was also the most extensive and complex procedure of its kind to date. The surgical team, which consisted of eleven surgeons, replaced virtually all of Stubblefield's facial tissue, including the scalp, eye sockets, nose, teeth, nerves, muscles, and skin. Before they did so, they paused so that a photographer could snap images of the disembodied face suspended between its two lives.

It was a landmark achievement. And it was the unwavering dedication of Harold Gillies and his team to the advancement of plastic surgery during the First World War that, ultimately, had made it possible for science fiction to become science fact.

NOTES

PROLOGUE: "AN UNLOVELY OBJECT"

3 *"I rather wondered"*: Private Papers of P. Clare, vol. 3, November 20, 1917. Private Papers of P. Clare. Documents. 15030. Documents and Sound Archives of the Imperial War Museums. The manuscript comprises four unpaginated volumes written in 1918, revised in 1920, and recopied in 1932 and 1935. It includes an appendix of transcribed letters to his mother.

4 *"They lay in trenches"*: William Clarke, "Random Recollections of '14/'18," 8, Liddle Collection, Brotherton Library Special Collections, University of Leeds. Originally found in Joanna Bourke, *Dismembering the Male: Men's Bodies, Britain, and the Great War* (London: Reaktion Books, 1996), 215.

4 *"they just shovelled everything"*: Quoted in Leo van Bergen, *Before My Helpless Sight: Suffering, Dying and Military Medicine on the Western Front, 1914–1918*, trans. Liz Waters (Farnham, UK: Ashgate, 2009), 490.

4 *"The worst," remembered one infantryman*: Robert Weldon Whalen, *Bitter Wounds: German Victims of the Great War, 1914–1939* (Ithaca, NY, and London: Cornell University Press, 1984), 43.

4 *A soldier could smell the front*: van Bergen, *Before My Helpless Sight*, 132.

4 *"Did you ever smell a dead mouse?"*: Paul Fussell, ed., *The Bloody Game: An Anthology of Modern Warfare*, vol. 2 (London: Abacus, 1992), 179.

4 *Clare had grown accustomed*: Quoted in Richard van Emden, *Meeting the Enemy: The Human Face of the Great War* (London: Bloomsbury, 2013), 186.

5 *"earth seemed to quake"*: Private Papers of P. Clare, vol. 3.

5 *"A few minutes later we moved on"*: Ibid.

6 *"[H]ow absurd it seemed"*: Ibid.

6 *Ammunition containing magnesium fuses*: Simon Schama, *The Face of Britain: The Nation Through Its Portraits* (London: Viking, 2015), 529.

7 *"[T]he science of healing"*: Ellen N. La Motte, "The Backwash of War: The Human Wreckage of the Battlefield as Witnessed by an American Hospital Nurse," in *Nurses at the Front: Writing the Wounds of the War*, ed. Margaret R. Higonnet (Boston: Northeastern University Press, 2001), 16.

7 *"They seemed to think"*: Fred H. Albee, *A Surgeon's Fight to Rebuild Men: An Autobiography* (New York: Dutton, 1945), 136.

7 *Some were even kicked*: Andrew Bamji notes that eleven patients who ended up at the Queen's Hospital in Sidcup received facial wounds inflicted by animals. Nine had been kicked, and two had been bitten. Andrew Bamji, *Faces from the Front: Harold Gillies, the Queen's Hospital, Sidcup and the Origins of Modern Plastic Surgery* (Solihull, UK: Helion, 2017), 21.

7 *Before the war was over*: Sandy Callister, "'Broken Gargoyles': The Photographic Representation of Severely Wounded New Zealand Soldiers," *Social History of Medicine* 20, no.1 (April 2007): 116–17; Suzannah Biernoff, "The Rhetoric of Disfigurement in First World War Britain," *Social History of Medicine* 24, no. 3 (January 2011): 666.

8 *"deploy firepower equivalent"*: van Bergen, *Before My Helpless Sight*, 31.

8 *"his face [was] black and charred"*: James William Davenport Seymour, *History of the American Field Service in France, "Friends of France" 1914–1917: Told by Its Members*, vol. 2 (Boston: Houghton Mifflin, 1920), 90.

8 *The first large-scale lethal gas attack*: The Germans tried gas a few times in 1914 on the Eastern Front, with no success.

8 *"Then there staggered into our midst"*: O. S. Watkins, *Methodist Report*, cited in Amos Fries and C. J. West, *Chemical Warfare* (New York: McGraw Hill, 1921), 13. Originally found in Gerard J. Fitzgerald, "Chemical Warfare and Medical Response During World War I," *American Journal of Public Health* 98, no. 4 (April 2008): 611–25.

9 *Like Percy Clare, Captain Jono Wilson*: Wilson, J. K. Tape 286/Transcript LIDDLE/WW1/TR/08/69, Liddle Collection, Brotherton Library Special Collections, University of Leeds.

9 *One pilot escaped with his body intact*: Frederick A. Pottle, *Stretchers: The Story of a Hospital Unit on the Western Front* (New Haven, CT: Yale University Press, 1929), chapter 4.

9 *These early airmen sometimes*: Nelson Wyatt, "First World War Flyers Risked Shortened Lifespan but Have Extended Legacy," *Canadian Press*, accessed October 8, 2020, http://ww1.canada.com/faces-of-war/first-world-war-flyers-risked-shortened-lifespan-but-have-extended-legacy.

10 *"It sank deep into his forehead"*: Sean Coughlan, "Graphic Eyewitness Somme Accounts Revealed," BBC News, November 17, 2016, accessed November 4, 2019, https://www.bbc.co.uk/news/education-37975358.

10 *Even discarded jam tins*: Andrew Robertshaw, *First World War Trenches* (Stroud, UK: History Press, 2014), 62.

11 *"there was not a sign of life"*: Reginald A. Colwill, *Through Hell to Victory: From Passchendaele to Mons with the 2nd Devons in 1918*, 2nd ed. (Torquay, UK: Reginald A. Colwill, 1927), 81–82.

11 *Between eight and ten million*: van Bergen, *Before My Helpless Sight*, 132.

11 *Whereas a missing leg*: Bourke, *Dismembering the Male*, 59.

11 *"knows that he can turn"*: "Worst Loss of All. Public Deeply Moved by War-Time Revelation," *Manchester Evening Chronicle*, May–June 1918. Found in the Queen's Hospital, Sidcup, Kent: Newspaper Cuttings, London Metropolitan Archives, City of London H02/QM/Y/01/005, page 37.

11 *"very severe facial disfigurement"*: Bourke, *Dismembering the Male*, 65. In contrast, disfigurement in France was a Class 6 injury, considered less serious than blindness or loss of limbs, and it therefore warranted no pension. As a result, the surgeon Léon Dufourmentel lamented that maimed French veterans were sure to face economic hardship: "It is sadly certain that a disfigured face inspiring disgust or horror, in spite of the pity and respect we owe to the victims of the Great War, can cause these men considerable prejudices." See Claudine Mitchell, "Facing Horror: Women's Work, Sculptural Practice and the Great War," in Valerie Mainz and Griselda Pollock, eds., *Work and the Image II: Work in Modern Times, Visual Mediations and Social Processes* (Aldershot, UK: Ashgate, 2000), 45.

12 *For centuries, a marked face*: Suzannah Biernoff, *Portraits of Violence: War and the Aesthetics of Disfigurement* (Ann Arbor: University of Michigan Press, 2017), 15.

12 *In fact, disfigurement carried with it*: Marjorie Gehrhardt, *The Men with Broken Faces*: Gueules Cassées *of the First World War* (Oxford: Peter Lang, 2015), 2. See also François-Xavier Long, "Les Blessés de la Face durant la Grande Guerre: Les Origines de la Chirurgie Maxillo-faciale," *Histoire des Sciences Médicales* 36, no. 2 (2002): 175–83.

12 *It can signify gender*: Biernoff, "The Rhetoric of Disfigurement," 669.

12 *The abrupt transformation*: Patricia Skinner, "'Better Off Dead Than Disfigured'? The Challenges of Facial Injury in the Premodern Past," *Transactions of the Royal Historical Society* 26 (2016): 26.

12 *"I supposed he [the doctor]"*: Francis J. McGowan, "My Personal Experiences of the Great War," page 7. 6 Mss Essays by Patients with Facial Injuries in Sidcup Hospital, 1922. LIDDLE/WW1/GA/WOU/34,

Essay 1. Liddle Collection, Brotherton Library Special Collections, University of Leeds.

13 *"victims of despondency"*: R. T. McKenzie, *Reclaiming the Maimed: A Handbook of Physical Therapy* (New York: Macmillan, 1918), 117.

13 *In Britain, they were known*: "The Loneliest of All Tommies," *Sunday Herald*, June 1918. Found in the Queen's Hospital, Sidcup, Kent: Newspaper Cuttings, London Metropolitan Archive, H02/QM/Y/01/005, page 41.

13 *"I had been through so many"*: Private Papers of P. Clare, vol. 3.

13 *His mind began drifting*: Ibid.

14 *"a load of iron chains"*: Ibid.

14 *"I imagined the burial parties"*: Ibid.

14 *He pulled a small Bible*: Letter from Percy Clare to his mother (n.d.), Private Papers of P. Clare, Letters to His Mother.

14 *A soldier named Ernest Wordsworth*: Ernest Wordsworth, "My Personal Experiences of the Great War," 6 Mss Essays by Patients with Facial Injuries in Sidcup Hospital, 1922. LIDDLE/WW1/GA/WOU/34, Essay 2. Liddle Collection, Brotherton Library Special Collections, University of Leeds.

14 *During the Battle of Loos*: van Bergen, *Before My Helpless Sight*, 306.

15 *"We stood there a moment"*: Louis Barthas, *Les Carnets de Guerre de Louis Barthas, Tonnelier 1914–1918* (Paris: Maspero, 1983), 72. Originally quoted in van Bergen, *Before My Helpless Sight*, 169–70.

15 *"Hello, Perc, poor old fellow"*: Private Papers of P. Clare, vol. 3.

15 *"I was so soaked with blood"*: Ibid.

16 *After Private W. Lugg*: Lyn MacDonald, *They Called It Passchendaele* (London: Michael Joseph, 1978), 118.

16 *"it was then just a question"*: Quoted in Ena Elsey, "Disabled Ex-Servicemen's Experiences of Rehabilitation and Employment After the First World War," *Oral History* 25, no. 2 (Autumn 1997): 51.

16 *"I was rendered speachless [sic]"*: "My Personal Experiences of the Great War," 6 Mss Essays by Patients with Facial Injuries in Sidcup Hospital, 1922. LIDDLE/WW1/GA/WOU/34, Essay 6. Liddle Collection, Brotherton Library Special Collections, University of Leeds.

16 *Early in the war*: Sir Harold Gillies and D. Ralph Millard, Jr., *The Principles and Art of Plastic Surgery* (London: Butterworth, 1957), 23.

17 *"I well remember wrapping him"*: Quoted in Sir Terence Ward, "The Maxillofacial Unit," *Annals of the Royal College of Surgeons of England* 57 (1975): 67. In 1962, William Kelsey Fry gave the opening address at the First International Congress of Oral Surgery.

17 *Only later in the war*: Notes on Maxillo-facial Injuries, report pre-

sented to Army Council, 1935, AWM54, 921/3/1. Found in Kerry Neale, "Without the Faces of Men: Facially Disfigured Great War Soldiers of Britain and the Dominions" (unpublished PhD thesis, University of New South Wales, Australia, March 2015), 47. Neale points out that it is difficult to know to what extent this protocol was adopted.

17 *He later referred to the wound*: Clare refers to the wound as a "Blighty One" in his diary, but the phrase is more often associated with less disabling injuries.

17 *"I was an unlovely object"*: Private Papers of P. Clare, vol. 3.

18 *"The operating theatre looked"*: Fritz August Voigt, *Combed Out* (London: Swarthmore Press, 1920), 70.

19 *Continuing improvements to this complex*: Bamji, *Faces from the Front*, 31.

1. THE BALLERINA'S RUMP

21 *With over seven million people*: Stefan Goebel and Jerry White, "London and the First World War," *London Journal* 41, no. 3 (2016): 199–218, accessed March 2, 2020, https://www.tandfonline.com/doi/full/10.1080/03058034.2016.1216758.

21 *London wasn't just big*: Ibid.

22 *The chief physician*: Reginald Pound, *Gillies, Surgeon Extraordinary: A Biography* (London: Michael Joseph, 1964), 15.

23 *"This is ridiculous"*: Ibid., 18.

23 *"Oh, my dear fellow"*: Ibid.

23 *"mysteriously inherited rather than"*: Ibid., 9.

24 *Early in life, Gillies fractured*: D. Ralph Millard. Jr., "Gillies Memorial Lecture: Jousting with the First Knight of Plastic Surgery," *British Journal of Plastic Surgery* 25 (1972): 73; Michael Felix Freshwater, "A Critical Comparison of Davis' *Principles of Plastic Surgery* with Gillies' *Plastic Surgery of the Face*," *Journal of Plastic, Reconstructive & Aesthetic Surgery* 64 (2011): 20. Freshwater indicates that there is some dispute about which elbow was fractured. Millard told Freshwater that it was the right elbow; Gillies's biographer Reginald Pound claimed that it was the left. A film of Gillies doing a forehead flap nasal reconstruction shows him hyperflexing his right wrist, which would be compatible with compensating for elbow stiffness.

24 *Two days before his fourth birthday*: "The Late Robert Gillies," *Bruce Herald*, Volume 17, Issue 1759, June 18, 1886, accessed April 20, 2020, https://paperspast.natlib.govt.nz/newspapers/BH18860618.2.12?fbclid=IwAR38tHrv_zbjP15-xLT5RJSa0MxgHqodP-4uIcrdRRils801XQnoRmOCqwU.

25 *"his happy temperament"*: Pound, *Gillies*, 14.

25 *In spite of his rebellious spirit*: Ibid., 15–16.

25 *It was during his studies*: Ibid., 13.

26 *"immense powers of concentration"*: Letter from Norman Jewson to Reginald Pound. Letters to Reginald Pound, 1955–1972. From the Archives of the Royal College of Surgeons, MS0336.

26 *Those who knew him*: Quoted in Pound, *Gillies*, 16.

26 *"Whatever he decided to do"*: Kenneth D. Pringle, Notes on Sir Harold Gillies. Letters to Reginald Pound.

27 *The stern etiquette of the day*: Mick Gillies, *Mayfly on the Stream of Time* (Whitfeld, UK: Messuage Books, 2000), 3.

27 *"but the slight lump"*: Pound, *Gillies*, 20.

29 *Lojka helped the archduke*: The date link first came to light when it was spotted by historian Brian Presland visiting the Museum of Military History in Vienna, where the car is on display, in 2004.

29 *The luxurious car*: Some accounts say that the archduke's car was third in the motorcade, but most scholars agree it was the second. Christopher Clark, *The Sleepwalkers: How Europe Went to War in 1914* (London: Penguin, 2013), 367–77. I'm hugely indebted to Clark for the details included in this section of the book.

30 *As chaos broke out*: Ibid., 371.

30 *"Come on. That fellow"*: Quoted ibid.

30 *"All of the citizens"*: Quoted ibid., 373.

30 *"I come here as your guest"*: Ibid.

31 *"Do you think Sarajevo"*: Greg King and Sue Woolmans, *The Assassination of the Archduke: Sarajevo 1914 and the Murder That Changed the World* (London: Pan Books, 2014), 204.

31 *"This is the wrong way!"*: Quoted in Clark, *Sleepwalkers*, 374.

31 *Lojka rolled to a stop*: Some accounts claim that the Gräf & Stift had no reverse gear. Mr. Ilming, an arms and technology expert at the Museum of Military History in Vienna, says the car does have a reverse gear but that it took a long time to switch gears because of the technical standards of the day. Benjamin Preston, "The Car That Witnessed the Spark of World War I," *New York Times*, July 10, 2014, accessed May 7, 2020, https://www.nytimes.com/2014/07/11/automobiles/the-car-that-witnessed-the-spark-of-world-war-i.html.

31 *Gavrilo Princip—who, like Čabrinović"*: Clark, *Sleepwalkers*, 375–76.

32 *Over two thousand articles*: This statistic comes from searching www.britishnewspaperarchive.co.uk.

32 *"Women at Boxing Matches"*: *Daily Record,* Saturday, June 27, 1914, 6.

33 *The attitude of Britain's politicians*: Matthew Johnson, "More Than Spectators? Britain's Liberal Government and the Decision to Go to

War in 1914," *The Conversation*, August 4, 2014, accessed February 5, 2019, http://theconversation.com/more-than-spectators-britains-liberal -government-and-the-decision-to-go-to-war-in-1914-30053.

34 "*hailed with wild, enthusiastic cheers*": *Daily Mirror*, August 5, 1914, 3.

34 "*it seemed as though the world*": "Voices of the First World War: Joining Up," Imperial War Museums podcast, accessed July 21, 2020, https://www.iwm.org.uk/history/voices-of-the-first-world-war -joining-up.

34 *Among these recruits*: Details gleaned from "The Teenage Soldiers of World War One," BBC News, November 11, 2014, accessed January 26, 2021, https://www.bbc.co.uk/news/magazine-29934965; see David Lister, *Die Hard, Aby! Abraham Bevistein—The Boy Soldier Shot to Encourage the Others* (Barnsley, UK: Pen & Sword Military, 2005).

34 *Their names were sometimes read aloud*: George Coppard, *With a Machine Gun to Cambrai: The Tale of a Young Tommy in Kitchener's Army, 1914–1918* (London: Her Majesty's Stationery Office, 1969), 19.

35 "*What a way to get leave*": Quoted in van Bergen, *Before My Helpless Sight*, 443–44.

35 *Back in Britain, the newly appointed Secretary of State*: Denis Winter, *Death's Men: Soldiers of the Great War* (London: Allen Lane, 1978), 23.

35 "*I was looking in a shop window*": "Voices of the First World War: Joining Up."

35 *Demuth—who had tried on several occasions*: Max Arthur, *Forgotten Voices of the Great War* (London: Random House, 2012), 18–22.

36 *By the end of 1915*: van Bergen, *Before My Helpless Sight*, 39.

36 "*I can see a girl now*": Quoted in Stella Bingham, *Ministering Angels* (London: Osprey Publishing, 1979), 132.

36 "*[t]he wounded came just as they were*": Ibid., 133.

37 *Claire Elise Tisdall, a volunteer nurse*: Lyn MacDonald, *The Roses of No Man's Land* (London: Michael Joseph, 1980), 165.

37 "*It is at a time like this*": Quoted in Janet S. K. Watson, "Wars in the Wards: The Social Construction of Medical Work in First World War Britain," *Journal of British Studies* 41 (October 2002): 493.

37 "*be imparted in a few bandaging classes*": Quoted ibid., 494.

37 "*My good lady, go home*": Quoted in Fiona Reid, *Medicine in First World War Europe: Soldiers, Medics, Pacifists* (London: Bloomsbury, 2017), 4.

38 *In January 1915*: Reginald Pound states that the Red Cross sent for Gillies at the end of January; Gillies says he went to France "early in 1915." Pound, *Gillies*, 22; Gillies and Millard, *Principles and Art of Plastic Surgery*, 6. See also "New Commandant of Belgian Field Hospital," *The Times*, May 5, 1915, 13.

38 *The decision to volunteer*: General Register Office, Margaret Gillies's birth registration, St. Marylebone, London (born January 31, 1915, 73 New Cavendish Street; registered February 25, 1915, by K. M. Gillies). PDF copy in possession of author.

2. THE SILVER GHOST

40 *Valadier spent most of his childhood*: William Cruse, "Auguste Charles Valadier: A Pioneer in Maxillofacial Surgery," *Military Medicine* 152, no. 7 (1987): 337–38.

41 *"The bastard! They made me pay"*: Quoted ibid., 338.

41 *However, rules stipulated*: Ibid., 338.

42 *During the Boer War*: M. J. Newell, ed., *Ex Dentibus Ensis: A History of the Army Dental Service* (Aldershot, UK: RADC Historical Museum, 1997), 11.

42 *Despite these hard lessons*: The army did eventually send dentists to South Africa.

42 *"Can a medical man say"*: "Dental Examiners to the Forces (Editorial)," *New Zealand Dental Journal* 10 (January 1915): 150–51. Found in Harvey Brown, *Pickerill: Pioneer in Plastic Surgery, Dental Education and Dental Research* (Dunedin, NZ: Otago University Press, 2007), 107.

42 *"numbers of men frankly stating"*: Quoted in Penny Starns, *Sisters of the Somme: True Stories from a First World War Field Hospital* (Stroud, UK: History Press, 2016), 55.

43 *"Man, ye're making a gran' mistake"*: *Punch* (August 19, 1914). Found in Nic Clarke, *Unwanted Warriors: Rejected Volunteers of the Canadian Expeditionary Force* (Vancouver: University of British Columbia Press, 2016), 82.

43 *"They must want blokes to bite"*: Robert Roberts, *The Classic Slum: Salford Life in the First Quarter of the Century* (Manchester, UK: Manchester University Press, 1971), 150.

43 *"two syringefuls [sic] of a solution"*: Starns, *Sisters of the Somme*, 13.

44 *"under a rain of bullets"*: Sylvestre Moreira, "The Dental Service in War," *Dental Surgeon* 14, no. 685 (December 15, 1917): 488, quoted in F.S.S. Gray, "The First Dentists Sent to the Western Front During the First World War," *British Dental Journal* 222, no. 11 (2017): 893.

44 *"excellent and most valuable surgical work"*: Quoted in Cruse, "Auguste Charles Valadier," 339. There is a question as to whether Valadier was the dentist who extracted General Haig's tooth. Cruse and other scholars have made a strong case for Valadier, given his proximity to the Battle of Aisne, his reputation, and the fact that Haig later recommended Valadier for a decoration. For these reasons, I have included this story in the book.

44 *By the end of 1914*: L. J. Godden, *History of the Royal Army Dental Corps* (Aldershot, UK: Royal Army Dental Corps, 1971), 5.

44 *This number gradually grew*: Ibid., 8.

45 *"the sheds were being converted"*: A. L. Walker, "A Base Hospital in France, 1914–1915," Scarletfinders, accessed May 18, 2020, http://www.scarletfinders.co.uk/156.html.

46 *Mortality rates from wound infections*: Neale, "Without the Faces of Men," 47.

46 *Ironically, the surgeon Joseph Lister*: For more on Joseph Lister, see Lindsey Fitzharris, *The Butchering Art: Joseph Lister's Quest to Transform the Grisly World of Victorian Medicine* (New York: Scientific American / Farrar, Straus and Giroux, 2017).

46 *"a war of faecal infection"*: A. G. Butler, *Official History of the Australian Medical Services, 1914–1918*, vol. 2 (Canberra: Australian War Memorial, 1940), 315.

46 *Even if antiseptic dressings*: Tom Scotland, *A Time to Die and a Time to Live, Disaster to Triumph: Groundbreaking Developments in Care of the Wounded on the Western Front 1914–18* (Warwick, UK: Helion, 2019), 71–77.

46 *Fortunately, Valadier recognized*: J. E. McAuley, "Charles Valadier: A Forgotten Pioneer in the Treatment of Jaw Injuries," *Proceedings of the Royal Society of Medicine* 67, no. 8 (1974): 786.

47 *He also offered his services*: Murray C. Meikle, *Reconstructing Faces: The Art and Wartime Surgery of Gillies, Pickerill, McIndoe and Mowlem* (Dunedin, NZ: Otago University Press, 2013), 48.

47 *It was at this specialist unit*: Gillies's exact movements in France are uncertain. I've pieced them together as best I could using a variety of sources cited in this chapter, but there are still gaps. In *The Principles and Art of Plastic Surgery*, Harold Gillies states that he arrived in France in early 1915. His biographer Reginald Pound corroborates this, stating it was the end of January 1915. An article in *The Times* states that Gillies arrived at the Belgian Field Hospital in Hoogstade in early May. The following month, he was promoted to major and reassigned to the Allied Forces' Base Hospital Étaples, France, where he remained until December. Before taking up his new post, he went on leave. He returned to London, where the sportswriter Henry Leach spotted him playing golf. According to Gillies, he traveled to Paris sometime in June to visit Hippolyte Morestin. Both Gillies and Pound imply that he met Valadier before he met Morestin. Therefore, it seems likely to me that Gillies met Valadier in the months *before* he was assigned to the Belgian Field Hospital. The period between January and May is otherwise unaccounted for in the records. It is possible that Gillies met Valadier after leaving the Belgian Field Hospital but before meeting Morestin. The timeline is so narrow, however, that I don't think this is a likely scenario. Records also state that Gillies was assigned to Valadier's jaw unit

to supervise the dentist's work, which implies it was an official post, rather than a chance encounter while he was on leave.

47 *The term "plastic surgery"*: One early use of the term in relation to surgery was by the German surgeon Carl Ferdinand von Graefe in his paper "Rhinoplastik," published in 1818. See John D. Holmes, "Development of Plastic Surgery," in *War Surgery 1914–18*, eds. Thomas Scotland and Steven Heys (Solihull, UK: Helion, 2012), 258.

48 *Known as a "minié ball"*: Pat Leonard, "The Bullet That Changed History," *New York Times*, August 31, 2012, accessed August 8, 2018, https:// opinionator.blogs.nytimes.com/2012/08/31/the-bullet-that-changed -history/.

48 *In one noteworthy case*: Gurdon Buck, *Case of Destruction of the Body of the Lower Jaw and Extensive Disfiguration of the Face from a Shell Wound* (Albany, NY: Private Printing, 1866), 4–5. Originally found in David Seed, Stephen C. Kenny, and Chris Williams, eds., *Life and Limb: Perspectives on the American Civil War* (Liverpool: Liverpool University Press, 2016), 90.

48 *One of Buck's most complicated cases*: The Medical and Surgical History of the War of the Rebellion (1861–65) / Prepared, in Accordance with the Acts of Congress, Under the Direction of Surgeon General Joseph K. Barnes, United States Army (Washington, DC: Government Printing Office, 1870–88), 721.

48 *Buck enlisted the help*: F. W. Seward, *Seward at Washington as Senator and Secretary of State: A Memoir of His Life, with Selections from His Letters (1861–1872)* (New York: Derby & Miller, 1891), 270. See also "Terrible Tragedy in Washington: Murder of the President, Attempted Murder of Mr. Seward," *New York Times*, April 17, 1865, 2.

48 *Buck then performed a series*: William W. Keen, "Gangrene of the Face Following Salivation," National Museum of Army Medicine Accession File 1000867; Gurdon Buck, *Contributions to Reparative Surgery* (New York: D. Appleton, 1876), 36, 38. Originally found in Seed, Kenny, and Williams, *Life and Limb*, 90–91.

49 *When Private Joseph Harvey*: George Alexander Otis, *The Medical and Surgical History of the War of the Rebellion*, part 1, vol. 2 (Washington, DC: Government Printing Office, 1876).

49 *As a result, fewer than forty*: Blair O. Rogers and Michael G. Rhode, "The First Civil War Photographs of Soldiers with Facial Wounds," *Journal of Aesthetic Plastic Surgery* 19 (1995): 271.

49 *"a charming, jaunty cowboy"*: Observation by Mr. C. Bowdler Henry in McAuley, "Charles Valadier," 788.

50 *"an impresario, a pretender"*: Quoted in J. E. McAuley, "Valadier Revisited," *Dental Historian* 19 (1990): 19.

50 *When Valadier presented Brigham*: Quoted ibid.

50 *"It was said that he was a millionaire"*: Garffild Lloyd Lewis, *Faced with Mametz* (Llanrwst, UK: Gwasg Carreg Gwalch, 2017), 89.

50 *In 1668, the Dutch surgeon*: Herman H. de Boer, "The History of Bone Grafts," *Clinical Orthopaedics and Related Research* 226 (January 1988): 292–98.

51 *In one case, Valadier*: Letter from Philip Thorpe to Reginald Pound, March 30, 1963, 2, Letters to Reginald Pound.

51 *"Firstly, he carved the shape"*: Lewis, *Faced with Mametz*, 92.

52 *Valadier wired the two ends*: Meikle, *Reconstructing Faces*, 49.

52 *"He suddenly put down the plate"*: Letter from Philip Thorpe to Mr. J. E. McAuley, May 29, 1965, Museum of Military Medicine, Aldershot, UK, RADCCF/3/3/4/55/VALA.

52 *Distraction osteogenesis only became*: Meikle, *Reconstructing Faces*, 52.

52 *"[t]he credit for establishing"*: Gillies and Millard, *Principles and Art of Plastic Surgery*, 6.

52 *"It was therefore inevitable"*: Ibid., 22.

53 *"I felt that I had not done"*: Harold Gillies, "The Problems of Facial Reconstruction," *Transactions of the Medical Society of London* 41 (1918): 165.

3. SPECIAL DUTY

55 *The Belgian Field Hospital*: The Times, May 5, 1915, 13. According to the newspaper, Gillies departed for Belgium with Morrison on Saturday, May 1, 1915.

55 *It was the hospital's third location*: Henry Sessions Souttar, *A Surgeon in Belgium* (London: Edward Arnold, 1915), 110–27.

55 *In January 1915*: C. P. Blacker, *Have You Forgotten Yet? The First World War Memoirs by C. P. Blacker* (Barnsley, UK: Leo Cooper, 2000), 17–18.

56 *Staff converted a dilapidated*: "Belgian Field Hospital," *The Times*, April 27, 1916, 8.

56 *Nurses, orderlies, and laundry maids*: *A War Nurse's Diary: Sketches from a Belgian Field Hospital* (New York: Macmillan, 1918), 86–88.

56 *"The Nearest Hospital to the Firing Line"*: The Times, September 7, 1915, 6.

56 *"Cows roamed around"*: A War Nurse's Diary, 87.

57 *"A great cesspool ran"*: Ibid.

57 *With him was Herbert W. Morrison*: The Times, May 5, 1915, 13.

57 *Unfortunately, the hospital staff*: Blacker, *Have You Forgotten Yet?*, 28–29.

58 *"For two or three weeks"*: A War Nurse's Diary, 102.

58 *"seize a mop and pail"*: Ibid., 99.

58 *It not only claimed tens of thousands*: van Bergen, *Before My Helpless Sight*, 65–66. The Germans introduced chlorine gas with minimal effect on the Russian Eastern Front at Bolimov earlier in the war, where it was so cold, the gas froze.

59 *"I could hardly believe my eyes"*: "Voices of the First World War: Gas Attack at Ypres," Imperial War Museums podcast, accessed May 6, 2020, https://www.iwm.org.uk/history/voices-of-the-first-world-war -gas-attack-at-ypres.

59 *"There they lay, fully sensible"*: *A War Nurse's Diary*, 99.

60 *In June 1915—not long after*: No. 7 was known as the "Allied Forces Base Hospital." It had a short existence with two hundred beds at the Hotel Christol, Boulogne, from October 23, 1914, to January 11, 1915. It reopened for about five months at Étaples from August to November 1915. Gillies took leave in late June and, on his return to France, stopped in Paris to visit Morestin. He likely arrived at the Allied Forces Base Hospital shortly before or just as it reopened in August 1915. William Grant Macpherson, *History of the Great War Based on Official Documents. Medical Services General History*, vol. 2 (London: His Majesty's Stationery Office, 1923), 73.

60 *Before taking up his new assignment*: Henry Leach, "The Golfer's Progress," *Illustrated Sporting and Dramatic News*, June 26, 1915, 470.

60 *"prepared to swear that recently"*: Ibid.

60 *The Germans were quick to establish*: Darryl Tong, Andrew Bamji, Tom Brooking, and Robert Love, "Plastic Kiwis—New Zealanders and the Development of a Specialty," *Journal of Military and Veterans' Health* 17 (October 2008): 12. See also W. H. Dolamore, "The Treatment in Germany of Gunshot Injuries of the Face and Jaws," *British Dental Journal* 37 (1916 War Supplement): 105–84.

60 *"I had struck a branch"*: Quoted in Pound, *Gillies*, 23. Pound claims it was a book that inspired Gillies. But Andrew Bamji later speculated that it may have been an article.

61 *"wild beast, swift and ferocious"*: Georges Duhamel, "A Muster of Ghosts," in *Light on My Days: An Autobiography*, trans. Basil Collier (London: J. M. Dent & Sons, 1948), 275.

61 *On another occasion, the surgeon*: Ibid., 276–77.

61 *"Few surgeons have ever shown"*: J. L. Faure, "H. Morestin (1869–1919)," *Presse Médicale* 27 (1919): 109. Originally found in Blair O. Rogers, "Hippolyte Morestin (1869–1919). Part I: A Brief Biography," *Aesthetic Plastic Surgery* 6 (1982): 143.

61 *In 1902, while he was studying medicine*: David Tolhurst, *Pioneers in Plastic Surgery* (Cham, Switzerland: Springer, 2015), 35–38. See also Rogers, "Hippolyte Morestin (1869–1919)," 141–47.

63 *"dagger-like sharpness"*: Gillies and Millard, *Principles and Art of Plastic Surgery*, 7.
63 *"I stood spellbound."*: Ibid.
63 *"it was the most thrilling thing"*: Ibid.
63 *A short while later*: Gillies remained at the Allied Forces Base Hospital until December 1915. *London Gazette,* Supplement 29415 (December 23, 1915), 12804.
64 *"[H]is use of weighted dentures"*: Gillies and Millard, *Principles and Art of Plastic Surgery*, 22.
64 *The two men would remain*: For more on the relationship between Gillies and Kazanjian, see Hagop Martin Deranian, *Miracle Man of the Western Front: Dr. Varaztad H. Kazanjian, Pioneer Plastic Surgeon* (Worcester, MA: Chandler House Press, 2007), 106–108.
64 *Gillies believed that if casualties*: Bamji, *Faces from the Front*, 48.

4. A STRANGE NEW ART

67 *"This is the very thing"*: Catherine Black, *King's Nurse—Beggar's Nurse* (London: Hurst & Blackett, 1939), 85.
67 *"He would not admit defeat"*: Ibid., 85–86.
68 *By the time Gillies*: Ibid., 84.
68 *"[A]ll those young men"*: La Motte, "The Backwash of War," 13–14.
68 *So severe was the shortage*: van Bergen, *Before My Helpless Sight*, 286–87.
69 *Although each man was supposed*: Bingham, *Ministering Angels*, 139. Possibly apocryphal.
69 *"mutilated debris in lieu of faces"*: Quoted in Gehrhardt, *The Men with Broken Faces*, 55.
69 *Mary Borden—a nurse*: Mary Borden, *The Forbidden Zone* (London: William Heinemann, 1929), 142.
69 *"You could not go through the horrors"*: Black, *King's Nurse—Beggar's Nurse,* 84.
69 *"My tongue was literally hanging out"*: Gillies and Millard, *Principles and Art of Plastic Surgery*, 8.
70 *"[w]e ought to have treated"*: Duhamel, "A Muster of Ghosts," 278.
70 *"I was told by the War Office"*: Gillies, letter to Millard, September 12, 1951. Quoted in Bamji, *Faces from the Front*, 49.
70 *Gillies persisted, knowing that*: Pound, *Gillies*, 25.
71 *"In no other part of the work"*: H. D. Gillies and L.A.B. King, "Mechanical Supports in Plastic Surgery," *The Lancet* (March 17, 1917): 412.
71 *"Disappointment is in store for him"*: H. D. Gillies, *Plastic Surgery of the Face Based on Selected Cases of War Injuries of the Face Including Burns* (London: Henry Frowde, 1920), 12.

72 *"Whenever the head of a careless soldier"*: Gillies and Millard, *Principles and Art of Plastic Surgery*, 11.

72 *"It seemed that the first important step"*: Ibid., 13.

72 *Before the war, surgeons*: Pound, *Gillies*, 26–27.

73 *The consequences of hasty, ill-conceived repairs*: Ibid., 40.

74 *Many cases presented unforeseen challenges*: Gillies and Millard, *Principles and Art of Plastic Surgery*, 12.

74 *"[The patient] has to lie face downward"*: E. D. Toland, *The Aftermath of Battle* (London: Macmillan, 1916), 43–44. Originally found in Neale, "Without the Faces of Men," 139.

75 *Therefore, at Aldershot, specific attention*: Lisa Haushoher, "Between Food and Medicine: Artificial Digestion, Sickness, and the Case of Benger's Food," *Journal of the History of Medicine and Allied Sciences* 73, no. 2 (2018): 169.

75 *As winter dragged on*: Mark Harrison, *The Medical War: British Military Medicine in the First World War* (Oxford: Oxford University Press, 2010), 10.

75 *"began to have a potential value"*: "Medical Work in the Field and at Home," chapter 66 of *The Times History of the War*, part 41, vol. 4 (London: The Times, 1915), 42.

75 *"as soon as the healing had occurred"*: Quoted in Pound, *Gillies*, 26.

75 *"Was it not all a dead-end occupation"*: La Motte, "The Backwash of War," 9.

76 *In an address to the Medical Society of London*: Gillies, "The Problems of Facial Reconstruction," 165–70.

76 *"A casualty was not a matter"*: Denis Winter, *Death's Men: Soldiers of the Great War* (London: Allen Lane, 1978), 83.

76 *"There could only be one bright spot"*: Albee, *A Surgeon's Fight to Rebuild Men*, 128.

76 *"Nothing could have been devised"*: "Experiments in Armour," *The Times*, July 22, 1915, 7.

78 *"we should require breast and body pieces"*: "Steel-Clad Soldiers," *The Times*, March 8, 1917, 3.

78 *Attempts to improve the helmet*: Neale, "Without the Faces of Men," 37.

78 *Those who knew him*: Pound, *Gillies*, 38, 49.

78 *"full of human kindness"*: Quoted ibid., 39.

78 *Similarly, Sergeant Reginald Evans*: Quoted ibid.

79 *"psychological effect on a man"*: Albee, *A Surgeon's Fight to Rebuild Men*, 110.

79 *"I thanked Heaven"*: Quoted in Pound, *Gillies*, 37.

79 *"I never felt any embarrassment"*: Ward Muir, *The Happy Hospital* (London: Simpkin, Marshall, 1918), 143. Originally found in Biernoff, "The Rhetoric of Disfigurement," 668.

79 *"a wholesome and pleasing specimen"*: Muir, *The Happy Hospital*, 143–44.

79 *"He would set to work"*: Black, *King's Nurse—Beggar's Nurse*, 86.

80 *"[T]he Great War in which millions"*: Ibid., 85.

80 *"This was a strange new art"*: Gillies and Millard, *Principles and Art of Plastic Surgery*, 10.

80 *The death of one Private William Henry Young*: Private William Henry Young dies in August 1916, but he is wounded in December 1915, which is why I've placed his story in this chapter.

80 *Young was a portly man*: Young's headstone records his age to be thirty-eight at the time of his death, but birth and death records indicate he was forty.

80 *That winter, Young was sent back*: Henry L. Kirby, *Private William Young V.C.: One of Preston's Heroes of the Great War* (Blackburn, UK: T.H.C.L. Books, 1985), 6–7.

81 *As dawn crept over the trenches*: Ibid., 9–10.

81 *Then, despite the severity of his own injuries*: "Private William Young VC," Lancashire Infantry Museum, accessed June 23, 2021, https://www.lancashireinfantrymuseum.org.uk/private-william-young-vc/.

81 *"I am sure that it must be a great consolation"*: Quoted in Kirby, *Private William Young V.C.*, 9.

81 *"I am naturally very proud"*: Quoted ibid., 10.

81 *A fund was established*: Ibid., 11–12.

82 *"I washed up and continued to chisel"*: Gillies and Millard, *Principles and Art of Plastic Surgery*, 25.

82 *"It was not a little disconcerting"*: Ibid.

83 *He told Gillies that the chloroform*: Kirby, *Private William Young V.C.*, 13.

83 *This effect was especially dangerous*: Scotland, *A Time to Die and a Time to Live*, 86.

83 *Having survived multiple gunshot wounds*: Kirby, *Private William Young V.C.*, 13.

83 *He left behind nine young children*: Ibid., 6.

83 *"utterly miserable" due to the fatal outcome*: "Death of the Preston V.C. After an Operation," *Yorkshire Post*, August 29, 1916, 6.

83 *"Everything that could be done"*: Ibid.

5. THE CHAMBER OF HORRORS

85 *It was the spring of 1916*: Gillies and Millard, *Principles and Art of Plastic Surgery*, 30.

86 *"majority of plastic operations"*: Gillies, *Plastic Surgery of the Face*, 23.

86 *"Surgical haste definitely led"*: Gillies and Millard, *Principles and Art of Plastic Surgery*, 30.

86 *"Nowhere does the sheer horror"*: "Moulding New Faces. From a Surgeon," *Daily Mail*, September 15, 1916, 3.

87 *"It did a great deal"*: Quoted in Pound, *Gillies*, 39–40.

87 *"Gillies had an air"*: Quoted ibid., 29.

87 *"great Henry Tonks"*: Quoted ibid.

87 *At six foot four inches*: Anthony Bertram, *Paul Nash: The Portrait of an Artist* (London: Faber and Faber, 1955), 39.

87 *"lean and ascetic looking"*: L. Morris, ed., *Henry Tonks and the "Art of Pure Drawing"* (Norwich, UK: Norwich School of Art Gallery, 1985), 8.

87 *The English painter Gilbert Spencer*: Gilbert Spencer, *Memoirs of a Painter* (London: Chatto & Windus, 1974), 31.

88 *"I used to feel the illness"*: Joseph Hone, *The Life of Tonks* (London: William Heinemann, 1939), 11.

88 *"I bring grave news"*: Ibid., 110.

88 *"Oh, it all makes the newspapers"*: Ibid.

88 *"his seriousness is depressing"*: Ibid.

89 *After the Russians*: Jerry White, *London in the Twentieth Century: A City and Its People* (London: Viking, 2001), 103–104.

89 *Germans made up 10 percent*: Goebel and White, "London and the First World War," 199–218, accessed March 2, 2020, https://www.tandf online.com/doi/full/10.1080/03058034.2016.1216758.

89 *"a crime against our Empire"*: Quoted in Richard Hough, *Louis and Victoria: The Family History of the Mountbattens*, 2nd ed. (London: Weidenfeld and Nicolson, 1984), 246.

89 *When Joseph Pottsmeyer lost his job*: *Hackney and Kingsland Gazette*, September 14, 1914. Originally found in Jerry White, *Zeppelin Nights: London in the First World War* (London: Bodley Head, 2014), 73.

89 *Another man, named John Pfeiffer*: *The Times*, September 26 and October 26, 1914.

89 *Anti-German sentiments*: White, *Zeppelin Nights*, 72.

90 *Dr. W. B. Cosens oversaw*: Hone, *Life of Tonks*, 111.

90 *In 1886, he took a post*: Suzannah Biernoff, "Flesh Poems: Henry Tonks and the Art of Surgery," *Visual Culture in Britain* 11, no. 1 (2010): 25.

91 *"the desire of art"*: George Moore, *Conversations in Ebury Street* (London: William Heinemann, 1924), 117.

91 *"I had no banking account"*: Henry Tonks, "Notes from 'Wander Years,'" *Artwork* 5 (1929): 235.

91 *"coming to-morrow"*: Quoted in Hone, *Life of Tonks*, 111.

91 *"I am leaving to-morrow"*: Quoted ibid., 112.

91 *"The wounds are horrible"*: Quoted ibid., 114–115.

92 *"I am not any use"*: Quoted ibid., 115.

92 *"A wisdom far above anything"*: Quoted ibid., 126.

92 *"I may have something"*: Quoted ibid.

92 *"the Duke of Wellington reduced"*: The comment is attributed to an unnamed "London hostess" by Pound, *Gillies*, 30.

92 *Tonks's anatomical training*: Ashley Ekins and Elizabeth Stewart, eds., *War Wounds: Medicine and the Trauma of Conflict* (Wollombi, NSW: Exisle Publishing, 2011), 69.

93 *"I am doing a number of pastel heads"*: Quoted in Hone, *Life of Tonks*, 127.

93 *"When I exhibit a picture"*: D. S. MacColl, "Professor Henry Tonks," *Burlington Magazine for Connoisseurs* 70 (February 1937): 94.

94 *"living damaged Greek head"*: Quoted in Hone, *Life of Tonks*, 127.

94 *"The whole visible horizon"*: "An Eyewitness. Remarkable Narrative," *Daily Mail*, Saturday, June 17, 1916, 4.

95 *"the guns go up in the air"*: Imperial War Museums Sound Recording: AC 4096, Reel 1 C. Falmer. Originally quoted in Nigel Steel and Peter Hart, *Jutland, 1916: Death in the Grey Wastes* (London: Cassell, 2003), 95.

95 *"There was a colossal double explosion"*: Imperial War Museums Documents: E. C. Cordeaux: Manuscript letter, ca. 6/1916. Originally quoted in Steel and Hart, *Jutland, 1916*, 95.

96 *"The [ship] was obliterated"*: Imperial War Museums Documents: Misc. 1010, R. Church Collection: S-King Hall, Typescript manuscript. Originally quoted in Steel and Hart, *Jutland, 1916*, 108.

96 *"I realised then"*: Quoted from "Jutland." Imperial War Museums podcast, accessed August 1, 2020, https://www.iwm.org.uk/history/voices -of-the-first-world-war-jutland.

96 *"There seems to be something wrong"*: A.E.M. Chatfield, *The Navy and Defence: The Autobiography of Admiral of the Fleet Lord Chatfield*, vol. 1 (London: William Heinemann, 1942), 143.

96 *Recognizing that his forces*: G. J. Meyer, *A World Undone: The Story of the Great War, 1914–1918* (New York: Bantam Books, 2015), 421.

97 *"I think they'd been killed"*: Quoted from "Jutland," Imperial War Museums podcast.

97 *"I . . . feel very different now"*: Ms. letter written and signed "Albert" by HRH Prince Albert, later King George VI, to Mrs. Eugenie Godfrey-Faussett, June 11, 1916, Imperial War Museums Documents, 2884. "Letter Written by HM King George VI Describing the Battle of Jutland, June 1916," accessed August 4, 2020, https://www.iwm.org.uk /collections/item/object/1030002834.

97 *An American journalist summed it up*: Juliet Gardiner and Neil Wenborn, eds., *The History Today Companion to British History* (London: Collins & Brown, 1995), 443.

97 *Back aboard the* Vanguard: "An Eyewitness. Remarkable Narrative," 4.

97 "*I read the muster*": Imperial War Museums Documents: C. Caslon Collection, "Recollections of the Battle of Jutland," 16. Quoted in Steel and Hart, *Jutland, 1916*, 402.

98 "*Human flesh had got into*": Quoted in Steel and Hart, *Jutland, 1916*, 415–16.

98 "*unrecognisable scraps of humanity*": Imperial War Museums Documents: D. Lorimer, Typescript, 122. Quoted in Steel and Hart, *Jutland, 1916*, 408.

98 "*It was an eerie scene*": Imperial War Museums Documents: Misc. 1010, R. Church Collection: J. Handley, Manuscript answer to questionnaire, ca. 1970–74. Quoted in Steel and Hart, *Jutland, 1916*, 409.

98 "*[T]hey were so overcome*": Imperial War Museums Documents: C. Caslon Collection, "Recollections of the Battle of Jutland," 17. Quoted in Steel and Hart, *Jutland, 1916*, 405.

98 "*We had only candle lamps*": Imperial War Museums Documents: C. E. Leake, Typescript from the original 1917 manuscript, 1971. Quoted in Steel and Hart, *Jutland, 1916*, 381–82.

98 "*They were so badly burnt*": Imperial War Museums Documents: G.E.D. Ellis, Microfilm copy of manuscript diary, 31.5/1916. Quoted in Steel and Hart, *Jutland, 1916*, 210.

99 "*A man will walk*": Imperial War Museums Documents: D. Lorimer: Typescript, 120–21. Quoted in Steel and Hart, *Jutland, 1916*, 160–61. Found in Bamji, *Faces from the Front*, 23.

99 "*quite unlike any burns*": Imperial War Museums Documents: D. Lorimer: Typescript, 120–21. Quoted in Steel and Hart, *Jutland, 1916*, 160–61. Found in Bamji, *Faces from the Front*, 23.

99 "*they die and die very rapidly*": Ibid., 24.

99 "*grim, weird, and ghoulish sight*": Imperial War Museums Documents: Misc. 1010, R. Church Collection: F. J. Arnold, Manuscript answer to questionnaire, ca. 1970–74. Quoted in Steel and Hart, *Jutland, 1916*, 159.

100 *Alexander MacLean, on board HMS* Lion: A. Maclean and H.E.R. Stephens, "Surgical Experiences in the Battle of Jutland," *Journal of the Royal Naval Medical Service* 2 (1916): 421–25.

100 *Unfortunately for many*: On July 18, 2009, the last surviving veteran of the battle, Henry Allingham, died at the age of 113, by which time he was the oldest documented man in the world and one of the last surviving veterans of the whole war. "Britain's Oldest Veteran Recalls WWI," BBC News, June 26, 2006, accessed July 28, 2021, http://news.bbc.co.uk/1/hi/uk/5098174.stm.

100 *Their injuries had rendered them*: Gillies, *Plastic Surgery of the Face*, 356.

100 "*How a man can survive*": Ibid.

100 *"There is hardly an operation"*: Ibid., 3–4.

101 *This began to change in the nineteenth century*: Thomas Dent Mütter, "Cases of Deformity from Burns, Relieved by Operations," *American Journal of the Medical Sciences*, n.s. 4 (1842): 66–80.

102 *Gillies wrote that his reconstructive work*: Gillies, *Plastic Surgery of the Face*, 4.

102 *"There are cases in which"*: Thomas Dent Mütter, *Cases of Deformity from Burns: Successfully Treated by Plastic Operations* (Philadelphia: Merrihew & Thompson, 1843), 17.

102 *He warned about the dangers*: Gillies, *Plastic Surgery of the Face*, 123.

102 *"In planning the restoration"*: Ibid., 8.

103 *"[W]ith long instruments he could fashion"*: Gillies and Millard, *Principles and Art of Plastic Surgery*, 11.

103 *"To my horror a great drop of pus"*: Ibid., 12.

103 *"Then I had to confess"*: Quoted in Pound, *Gillies*, 36.

103 *Before leaving, he told Gillies*: Ibid., 33.

6. THE MIRRORLESS WARD

105 *"We were informed by all officers"*: Private Papers of S. W. Appleyard, Imperial War Museums 7990, 82/1/1 52–60. Found in Andrew Roberts, *Elegy: The First Day on the Somme* (London: Head of Zeus, 2015), 86.

106 *"When we started to fire"*: Quoted in G. J. Meyer, *A World Undone: The Story of the Great War, 1914–1918* (New York: Bantam Books, 2015), 386.

106 *"a veritable inferno"*: Private Papers of Major A. E. Bundy, Imperial War Museums Documents, 10828. Quoted in Anthony Richards, *The Somme: A Visual History* (London: Imperial War Museums, 2016), 79.

106 *Seymour had only advanced*: Gillies and Millard, *Principles and Art of Plastic Surgery*, 17–18; Gillies, *Plastic Surgery of the Face*, 264–65. Seymour's story is also discussed in Bamji, *Faces from the Front*, 19, 77, 172, and 192.

107 *"There were men everywhere"*: Quoted in van Bergen, *Before My Helpless Sight*, 83.

107 *"men on the barbed wire"*: "Voices of the First World War: The First Day of the Somme," Imperial War Museums podcast, accessed August 6, 2019, https://www.iwm.org.uk/history/voices-of-the-first-world-war-the-first-day-of-the-somme.

107 *"It seemed to me that everyone"*: Sean Coughlan, "Graphic Eyewitness Somme Accounts Revealed," BBC News, November 17, 2016, accessed August 3, 2020, https://www.bbc.co.uk/news/education-37975358.

107 *Of the 100,000 British soldiers*: van Bergen, *Before My Helpless Sight*, 77.

107 *Never before or since has a single army*: Ibid.

107 *This was in stark contrast*: Martin Middlebrook, *The First Day on the Somme* (Barnsley, UK: Pen & Sword Military, 2003 reprint), 264. Middlebrook notes that the exact German losses for the day's fighting cannot be known, as their units only made a casualty return once every ten days.

108 *They spilled out onto the grounds*: Starns, *Sisters of the Somme*, 94.

108 *"We had so many wounded"*: Quoted in Elsey, "Disabled Ex-Servicemen's Experiences of Rehabilitation and Employment After the First World War," 52.

108 *"Every so often the surgeon"*: Ibid.

108 *Philip Gibbs, who served*: Philip Gibbs, *Realities of War* (London: Heinemann, 1920), 287.

109 *"Let us roll up our sleeves"*: Gillies and Millard, *Principles and Art of Plastic Surgery*, 12.

109 *"There were wounds far worse"*: Quoted in Pound, *Gillies*, 33.

109 *"In all my nursing experience"*: Black, *King's Nurse—Beggar's Nurse*, 87.

110 *A similar technique appeared in Europe*: Whether this technique was independently conceived by Gustavo or reached Italy from India is debatable. For more information, see Isabella C. Mazzola and Riccardo F. Mazzola, "History of Reconstructive Rhinoplasty," *Journal of Facial Plastic Surgery* 30, no. 3 (2014): 227–36.

110 *This new technique involved*: Ambrose Paré, *The Workes of that famous Chirurgion Ambrose Parey, Translated out of Latine and compared with the French by Th[omas] Johnson* (London: Printed by Th. Cotes and R. Young, 1634), sig. Ddd4(v). Originally found in Emily Cock, "'Lead[ing] 'Em by the Nose into Publick Shame and Derision': Gaspare Tagliacozzi, Alexander Read and the Lost History of Plastic Surgery, 1600–1800," *Social History of Medicine* 28, no. 1 (2015): 7.

110 *"so resembling nature's pattern"*: Gaspare Tagliacozzi, "Letter to Mercuriale," in Martha Teach Gnudi and Jerome Pierce Webster, *The Life and Times of Gaspare Tagliacozzi: Surgeon of Bologna, 1545–1599* (New York: Rechner, 1950), 137.

111 *In 1705, the satirical writer*: Edward Ward and Thomas Brown, *The Legacy for the Ladies: Or, Characters of the Women of the Age* (London: S. Briscoe, 1705), sig. M4(v). Originally cited in Cock, "'Lead[ing] 'Em by the Nose,'" 2.

111 *Indeed, the fear of nasal deformity*: Cock, "'Lead[ing] 'Em by the Nose,'" 2.

111 *"restore, repair, and make whole"*: Quoted in Sander L. Gilman, *Making the Body Beautiful* (Princeton, NJ: Princeton University Press, 1999), 68.

111 *"some wine was warmed"*: René-Jacques Croissant de Garengeot, *Traité des Opérations de Chirurgie* (Paris: Huart, 1731), 55. Quoted in Thomas

Gibson, "Early Free Grafting: The Restitution of Parts Completely Separated from the Body," *British Journal of Plastic Surgery* 18 (1965): 3.

112 *"a Spaniard [named] Andreas Gutiero"*: J. Thomson, *Lectures on Inflammation* (Edinburgh: William Blackwood, 1813), 230.

112 *The historian Sander Gilman*: Gilman, *Making the Body Beautiful*, 71.

112 *Even though rhinoplasty had been around*: Bamji, *Faces from the Front*, 75.

113 *One patient who found himself*: Pound refers to this case briefly but claims that the patient had undergone surgery while in German captivity. Bamji expands upon this and identifies him as Leonard Tringham—who in actuality had undergone this failed operation in Birmingham. Pound, *Gillies*, 55; Bamji, *Faces from the Front*, 78; Gillies, *Plastic Surgery of the Face*, 228.

113 *Soon after Private Seymour's own arrival*: Gillies and Millard, *Principles and Art of Plastic Surgery*, 17; Gillies, *Plastic Surgery of the Face*, 264–65.

113 *William Spreckley—the eldest son of a lace-maker*: Kelly Smale, "William Spreckley Was Treated at Queens Hospital in Sidcup in 1917," *News Shopper*, July 19, 2012, accessed January 8, 2020, https://www.newsshopper.co.uk/news/9826677.william-spreckley-was-treated-at-queens-hospital-in-sidcup-in-1917/.

114 *Gillies wasted no time*: Gillies, *Plastic Surgery of the Face*, 294–98.

114 *Gillies likened it to an anteater's snout*: Gillies and Millard, *Principles and Art of Plastic Surgery*, 40.

115 *"[H]asty judgment leads often"*: Ibid.

115 *"Look at Spreckley today"*: Ibid.

115 *"[W]ithout it I was lost"*: Quoted in Pound, *Gillies*, 37.

115 *"All the time, we were fumbling towards"*: Quoted ibid., 39.

115 *"that silent ward where only one"*: Black, *King's Nurse—Beggar's Nurse*, 86–87.

115 *Gillies wondered if some of his patients*: Quoted in Pound, *Gillies*, 23.

116 *"When I first got a chance to examine myself"*: Capt. Holtzapffel, "Amateur Soldier" (unpublished and undated), 86, Liddle Collection, Brotherton Library Special Collections, University of Leeds, G. A. Wounds 58. Originally found in Fiona Reid, "Losing Face: Trauma and Maxillofacial Injury in the First World War," in Jason Crouthamel and Peter Leese (eds.), *Psychological Trauma and the Legacies of the First World War* (Basingstoke, UK: Palgrave Macmillan, 2017), 32.

116 *"If our plastic plans went wrong"*: Quoted in Pound, *Gillies*, 35.

116 *Preventing men from catching sight*: Black, *King's Nurse—Beggar's Nurse*, 87–89.

118 *"[n]othing was more painful"*: W. Arbuthnot Lane, "War." Unpublished Autobiography. Wellcome Library, Archives and manuscripts GC/127/A/2.

118 *When at last the young man*: Pound, *Gillies*, 35. Nurse Black does not mention this detail in her account. This comes from a direct quote by Gillies to Pound about the fate of this particular patient.
119 *Shortly after the Somme offensive*: The story of Private Walter Ashworth is discussed extensively in Schama, *The Face of Britain*, 532–34.
119 *"Unfortunately the missiles"*: Gillies and Millard, *Principles and Art of Plastic Surgery*, 13.
120 *If "facial hunks" were missing*: Bamji, *Faces from the Front*, 72–73.
120 *Ashworth underwent three painful operations*: Walter Ashworth case notes. From the Archives of the Royal College of Surgeons, British Patient Files MS0513/1/1/01 (54).
120 *"whimsical, one-sided expression"*: Gillies, *Plastic Surgery of the Face*, 62.
120 *Unfortunately, Ashworth's fiancée*: Bamji, *Faces from the Front*, 185. Bamji corresponded with Ashworth's grandaughter Diane Smith on this story.
121 *Once Ashworth was discharged*: Gillies, *Plastic Surgery of the Face*, 62.
121 *When he returned to his old job*: Emma Clayton, "Pioneering Plastic Surgery for Soldier Shot in the Face and Left for Dead in a Trench," *Telegraph & Argus*, November 7, 2018, accessed July 15, 2019, https://www.thetelegraphandargus.co.uk/news/17206411.pioneering-plastic-surgery-for-soldier-shot-in-the-face-and-left-for-dead-in-a-trench/.
121 *According to his granddaughter*: Ibid. There are conflicting accounts as to whether Ashworth underwent further surgery. I have chosen to base my account on the recollections of his granddaughter Diane Smith.

7. TIN NOSES AND STEEL HEARTS

123 *Like Tonks, Wood was too old*: Sarah Crellin, "Wood, Francis Derwent (1871–1926), sculptor," *Oxford Dictionary of National Biography* (September 23, 2004), accessed May 18, 2020, https://www.oxforddnb.com/view/10.1093/ref:odnb/9780198614128.001.0001/odnb-9780198614128-e-36999/version/1.
124 *"My work begins where the work"*: Francis Derwent Wood, "Masks for Facial Wounds," *The Lancet* (June 23, 1917): 949.
124 *Disease also played a role*: Brian F. Conroy, "A Brief Sortie into the History of Cranio-oculofacial Prosthetics," *Facial Plastic Surgery* 9, no. 2 (1993): 100.
124 *A book by the sixteenth-century military surgeon*: Meikle, *Reconstructing Faces*, 64.
124 *When Private Alphonse Louis*: G. Whymper, "The Gunner with the Silver Mask," *London Medical Gazette* 12 (1832–33): 705–709. See also, Conroy, "A Brief Sortie into the History of Cranio-oculofacial Prosthet-

ics," 89–115. The mask and plaster cast are now housed at Surgeons' Hall Museums in Edinburgh.

126 *"As they were in life"*: Wood, "Masks for Facial Wounds," 949.

126 *In contrast, Gillies was comfortable*: Letter from Horace Sewell to Reginald Pound, March 17, 1963, 4, Letters to Reginald Pound.

126 *Unlike artificial limbs*: Katherine Feo, "Invisibility: Memory, Masks and Masculinities in the Great War," *Journal of Design History* 20 (2007): 22.

126 *The production was tedious*: Sharon Romm and Judith Zacher, "Anna Coleman Ladd: Maker of Masks for the Facially Mutilated," *Plastic and Reconstructive Surgery* 70, no. 1 (1982): 108.

126 *If the changes to the underlying tissue*: Wood, "Masks for Facial Wounds," 949–51.

126 *"rob war of its ultimate horror"*: "Mending the Broken Soldier," *The Times*, August 12, 1916, 9.

126 *"The patient acquires his old self-respect"*: Wood, "Masks for Facial Wounds," 949.

127 *Each Tuesday, Ladd arranged*: Conroy, "A Brief Sortie into the History of Cranio-oculofacial Prosthetics," 105.

128 *"They were never treated"*: Muriel Caswall, "Woman Who Remade Soldiers' Injured Faces," *Boston Sunday Post*, February 16, 1919. Found in David M. Lubin, "Masks, Mutilation, and Modernity: Anna Coleman Ladd and the First World War," *Archives of American Art Journal* 47, no. 3/4 (2008): 10.

128 *"hideous flabby heap [for a nose]"*: Quoted in Higonnet, ed., *Nurses at the Front*, 63.

128 *"One man who came to us"*: "Finds Soldiers Brave Under Disfigurement," *Evening Public Ledger,* March 7, 1919, 11. See also Julie M. Powell, "About-Face: Gender, Disfigurement and the Politics of French Reconstruction, 1918–24," *Gender & History* 28, no. 3 (November 2016): 604 22.

129 *Together, they made ninety-seven masks*: Romm and Zacher, "Anna Coleman Ladd," 108. After Ladd left France, another person took over as director of the studio, and it continued to produce masks for disfigured men for another year before closing.

129 *Each one sold for eighteen dollars*: G. S. Harper. "New Faces for Mutilated Soldiers," *Red Cross Magazine* 13, no. 44 (November 1918).

129 *"They looked for all the world"*: Muriel Caswall, "Woman Who Remade Soldiers' Injured Faces Reaches Boston Home," *Boston Sunday Post*, February 16, 1919. Clipping in the Anna Coleman Ladd Papers, AAA, Box 2, Scrapbook, 1914–1923 (folder 4 of 7). Found in Biernoff, *Portraits of Violence*, 105.

129 *A recipient donning one of Ladd's masks*: Romm and Zacher, "Anna Coleman Ladd," 109.

129 *"sat like clods in the hospitals"*: "Her War Work Brings Honors," interview by Elizabeth Borton of the *Boston Herald*, November 29, 1932, Box 3, Folder 32, Anna Coleman Ladd papers, 1881–1950, Archives of American Art, Smithsonian Institution.

129 *As the war progressed*: Conroy, "A Brief Sortie into the History of Cranio-oculofacial Prosthetics," 106.

130 *Yet, for Gillies, the mere existence*: Biernoff, "The Rhetoric of Disfigurement," 666–85.

130 *"the cases that come to me"*: Wood, "Masks for Facial Wounds," 949.

130 *"collapsed in a faint"*: Letter from Frances Steggall to Reginald Pound, Letters to Reginald Pound.

131 *When one of Wood's patients*: Pound, *Gillies*, 35.

131 *"One can appreciate a sweetheart's repugnance"*: Quoted ibid., 50.

131 *"These blankety tin faces"*: H. P. Pickerill, "The Queen's Hospital, Sidcup," *British Journal of Plastic Surgery* 6 (1953): 249.

131 *"once we started their repair"*: Gillies and Millard, *Principles and Art of Plastic Surgery*, 10.

132 *"My days and nights"*: Quoted in Pound, *Gillies*, 38.

132 *"I found myself operating"*: Gillies and Millard, *Principles and Art of Plastic Surgery*, 30.

132 *This would free up beds*: Bamji, *Faces from the Front*, 52.

133 *Gillies badgered the "brass hats"*: Gillies and Millard, *Principles and Art of Plastic Surgery*, 30.

133 *The newly formed committee*: Bamji, *Faces from the Front*, 53–54.

134 *"I want to make Sidcup"*: Quoted in Pound, *Gillies*, 42.

134 *"one of the saddest conditions"*: Black, *King's Nurse—Beggar's Nurse*, 92.

134 *At the beginning of hostilities*: Ben Shephard, *A War of Nerves: Soldiers and Psychiatrists in the Twentieth Century* (Cambridge, MA: Harvard University Press, 2001), 21. See also Adam Montgomery, *The Invisible Injured: Psychological Trauma in the Canadian Military from the First World War to Afghanistan* (Montreal: McGill-Queen's University Press, 2017), 31–32.

134 *The writer Reginald Pound*: Pound, *Gillies*, 41.

135 *It was there that many*: Meikle, *Reconstructing Faces*, 81.

135 *"We literally put down our suitcases"*: Gillies and Millard, *Principles and Art of Plastic Surgery*, 31.

8. THE MIRACLE WORKERS

137 *"These spots here are the eyes"*: Harold Begbie, "Patient's New Face Taken from His Chest," *Yorkshire Evening Post*, December 6, 1917, 1. Beg-

NOTES

bie visits the Queen's Hospital in December 1917; this event is, therefore, out of sequence with patients yet to be discussed. I've taken artistic license and moved a discussion of his visit here, as it does not affect the historical integrity of the story.

137 *"War is horrible, devilish"*: Harold Begbie, "The Workshops of Destruction: Things Seen Behind the Firing Line," *Liverpool Daily Post*, March 27, 1915, 4.

138 *"[T]his is where the nose will go"*: Begbie, "Patient's New Face Taken from His Chest," 1.

138 *"I can see the patient is a man"*: Ibid.

138 *"[t]hat pencilled face on the man's chest"*: Ibid.

138 *"You see those little swellings"*: Ibid.

139 *"[T]he whole face on the chest"*: Ibid.

139 *"Mr. Derwent Wood, the most imaginative"*: Ibid.

139 *"A revolution has come"*: Ibid.

139 *Looking at the photos*: Ibid.

140 *"On purpose [the wound] is sewn up clumsily"*: Quoted in Richard Hopton, *Pistols at Dawn: A History of Duelling* (London: Portrait, 2007), 357.

140 *Young Germans sought out*: Gilman, *Making the Body Beautiful*, 123.

140 *He wanted to be able to "pass"*: Kun Hwang, "An Honorable Scar on the Face: A Scar Worthy of Satisfaction," *Journal of Craniofacial Surgery* 29, no. 8 (November 2018): 2009.

140 *Unlike Gillies, Joseph had experience*: Surajit Bhattacharya, "Jacques Joseph: Father of Modern Aesthetic Surgery," *Indian Journal of Plastic Surgery* 41 (October 2008): S3–S8.

141 *Afterward, Joseph joined a private practice*: Neale, "Without the Faces of Men," 88.

141 *Joseph won acclaim after he reconstructed*: Paolo Santoni-Rugiu and Philip J. Sykes, *A History of Plastic Surgery* (Berlin and London: Springer, 2007), 313.

141 *When surgeons visited the Charité*: Ibid., 313.

142 *"The larger the hospital"*: Pickerill, "The Queen's Hospital, Sidcup," 247–49.

142 *Gillies had organized the hospital*: Gillies and Millard, *Principles and Art of Plastic Surgery*, 30–31.

142 *When the journalist Harold Begbie visited*: Ibid., 30. Andrew Bamji puts this figure closer to seven hundred. Bamji, *Faces from the Front*, 64.

143 *"[N]ot until the organisation"*: Gillies, *Plastic Surgery of the Face*, ix.

143 *Early in the war*: Gillies and Millard, *Principles and Art of Plastic Surgery*, 23–24.

143 *The British section was the largest*: Pickerill, "The Queen's Hospital, Sidcup," 247.

143 *"He was a disciplinarian"*: Daryl Lindsay, "Five Men," *Medical Journal of Australia* (January 18, 1958): 62. Originally found in Bamji, *Faces from the Front*, 62. Lindsay first met Newland in France in 1916. Little did he know then that he would end up working for the surgeon for two years when he arrived at Sidcup.

144 *Next, there was Major Carl Waldron*: Meikle, *Reconstructing Faces*, 79. See also Bamji, *Faces from the Front*, 61–62.

144 *"all medical and surgical cases"*: H. P. Pickerill, "New Zealand Expeditionary Force, Jaw Department," *NZDJ* 13 (September 1917): 35–38. Originally quoted in Brown, *Pickerill*, 120.

144 *Over time, he began to specialize*: Meikle, *Reconstructing Faces*, 92.

145 *"Well, Major, to please me"*: HM Queen Mary, *Requiescat in Pace et Honore*. Unpublished and undated manuscript. Pickerill Papers, Hocken Collections. Originally quoted in Brown, *Pickerill*, 123.

146 *"I had been sent to France"*: Quoted in Deranian, *Miracle Man of the Western Front*, 106–107.

146 *"As the years of the War continued"*: Ibid.

146 *"This was indeed an impressive array"*: Gillies and Millard, *Principles and Art of Plastic Surgery*, 31.

146 *"The whole hospital was an excellent example"*: Pickerill, "The Queen's Hospital," 249.

147 *"Such an intensive culture"*: "Intensive Medical Treatment," *The Lancet* (December 8, 1917): 863.

147 *"With our artistic efforts"*: Gillies and Millard, *Principles and Art of Plastic Surgery*, 38.

147 *"competition brought out many men"*: T. B. Layton, *Sir William Arbuthnot Lane, Bt. C.B., M.S.: An Enquiry into the Mind and Influence of a Surgeon* (Edinburgh and London: E. & S. Livingstone, 1956), 111.

147 *Medical patents on surgical techniques*: Sally Frampton, "Honour and Subsistence: Invention, Credit and Surgery in the Nineteenth Century," *British Journal for the History of Science* 49 (December 2016): 566.

147 *In such an environment*: Gillies, *Plastic Surgery of the Face*, x.

148 *"Reprint extracted and sent to Gillies"*: Quoted in Brown, *Pickerill*, 129.

148 *"How a man can survive"*: Gillies, *Plastic Surgery of the Face*, 356.

148 *"it required very considerable moral courage"*: Ibid.

148 *"unquenchable optimism which carries them"*: Ibid.

150 *"If I stitched the edges"*: Quoted in Pound, *Gillies*, 44.

150 *"[A]nother needle was threaded"*: Quoted ibid., 45.

150 *"Those tubes of Seaman Vicarage"*: Quoted ibid.

151 *"I could bring them from one part"*: Quoted ibid.

151 *"I could make them in the form"*: Quoted ibid.

151 *"The enthusiastic rabble of surgeons"*: Gillies and Millard, *Principles and Art of Plastic Surgery*, 37.

151 *Ten years after the war*: Pound, *Gillies*, 78.

151 *"If all the tube pedicles"*: Gillies and Millard, *Principles and Art of Plastic Surgery*, 153.

152 *"I don't mind reading"*: Ibid., 37.

152 *"As in all innovations, limitations"*: Ibid.

152 *One evening, Gillies picked his way*: Pound, *Gillies*, 46.

152 *In East London, a bomb*: "World War One: How the German Zeppelin Wrought Terror," BBC News, August 4, 2014, accessed February 24, 2020, https://www.bbc.co.uk/news/uk-england-27517166.

153 *Doris Cobban wrote of being five years old*: Ibid.

153 *"You had about as much chance"*: Quoted in Patrick Bishop, *Wings: One Hundred Years of British Aerial Warfare* (London: Atlantic Books, 2012), 82.

153 *Still, some pilots got lucky*: Thomas Fegan, *The "Baby Killers": German Air Raids on Britain in the First World War* (Barnsley, UK: Pen & Sword Military, 2013), 21–22.

154 *"The Kaiser, God damn him"*: Albee, *A Surgeon's Fight to Rebuild Men*, 85.

154 *Alongside outrage, however, there was also alarm*: Christopher Klein, "London's World War I Zeppelin Terror," *History* (August 31, 2018), accessed February 25, 2020, https://www.history.com/news/londons -world-war-i-zeppelin-terror.

154 *On September 2, 1916*: Ibid.

155 *"It had broken in half"*: Ms. letter from Patrick Blundstone to his father, September 1916, Imperial War Museums Documents 5508, "Letter Concerning the Burning of a Zeppelin," accessed August 4, 2020, https://www.iwm.org.uk/collections/item/object/1030005513.

155 *The airships had betrayed*: Christopher Cole and E. F. Cheesman, *The Air Defence of Great Britain 1914–1918* (London: Putnam, 1984), 448– 49; Micheal Clodfelter, *Warfare and Armed Conflicts: A Statistical Encyclopedia of Casualty and Other Figures, 1492–2015* (Jefferson, NC: McFarland, 2017), 430. According to Cole and Cheesman, 557 people were killed and 1,358 were injured by Zeppelins. According to Clodfelter, 857 people were killed and 2,058 were injured by the heavier biplane bombers. I arrived at the total number of casualties by adding these numbers together.

155 *The number of German civilians*: Jay Winter and Jean-Louis Robert, *Capital Cities at War: Paris, London, Berlin 1914–1919*, vol. 1 (Cambridge: Cambridge University Press, 1997), 517.

155 *"The poor chap had to sleep"*: Quoted in Pound, *Gillies*, 46.

156 *He knew of a technique*: B. Haeseker, "The First Anglo-Dutch Con-
tacts in Plastic Surgery: A Brief Historical Note," *British Journal of Plastic
Surgery* 38 (1985): 15–23.

156 *In 1916, Esser took a skin graft*: J.F.S. Esser, "Epithelial Inlay in Cases
of Refractory Ectropion," *Archives of Ophthalmology* 16, no. 1 (1936):
55–57.

156 *He published his findings first in German*: J. F. Esser, "Studies in Plastic Surgery of the Face," *Annals of Surgery* 65, no. 3 (March 1917):
297–315.

156 *Several weeks after Vicarage's first operation*: For more detailed descriptions of the epithelial outlay, see Meikle, *Reconstructing Faces*, 83–84.

157 *There was "doubting interest"*: Pound, *Gillies*, 46.

157 *Indeed, the epithelial outlay*: Bamji, *Faces from the Front*, 95.

9. THE BOYS ON BLUE BENCHES

159 *Doris Maud was only eleven*: John Mercer, *Sidcup & Foots Cray: A History*
(Stroud, UK: Amberley Publishing, 2013), 52.

160 *"What kind of vision"*: "Miracles They Work at Frognal," *Daily Sketch*,
April 1918. Found in Biernoff, *Portraits of Violence*, 18–19.

160 *"rudest blow that war can deal"*: "Worst Loss of All."

160 *"needn't have the slightest worry"*: Letter dated March 10, 1916. GS
1816, Evans, Reginald, J. T., box 1. Liddle Collection, Brotherton
Library Special Collections, University of Leeds. Found in Biernoff,
Portraits of Violence, 68.

160 *"to cultivate these good looks"*: Quoted ibid., 68–69.

161 *"will have to prepare"*: Ibid.

161 *"You wait till I come swanking home"*: Ibid.

161 *Despite the public's reluctance*: Bamji, *Faces from the Front*, 143.

161 *"He is aware of just what he looks like"*: Muir, *The Happy Hospital*, 144.

161 *"This, then, is the patient"*: Ibid.

162 *"The good people of that place"*: Letter from Horace Sewell to Reginald Pound, March 17, 1963, 2, Letters to Reginald Pound.

162 *"he had his way"*: Ibid., 3.

162 *Harold Gillies stood in the alleyway*: Pound, *Gillies*, 47.

163 *On one occasion, Gillies managed to swap*: G. M. FitzGibbon, "The
Commandments of Gillies," The Gillies Lecture 1967, *British Journal of
Plastic Surgery* 21 (1968): 227.

163 *"We played our game"*: Quoted in Pound, *Gillies*, 47.

163 *"Major Gillies himself was"*: Letter from Philip Thorpe to Reginald
Pound, March 11, 1963, 6, Letters to Reginald Pound.

163 *"The injury to the subconscious mind"*: Quoted in Pound, *Gillies*, 50.

164 *"We noticed that if we made"*: Gillies and Millard, *The Principles and Art of Plastic Surgery*, 45.

164 *"Two more eggs, two rounds of toast"*: Letter from Philip Thorpe to Reginald Pound, March 11, 1963, 7, Letters to Reginald Pound.

164 *"I can still look back"*: Budd Papers, Liddle Collection, Brotherton Library Special Collections, University of Leeds, LIDDLE/WWI/WF/REC/01/B43. Quoted in Bamji, *Faces from the Front*, 148.

165 *"However," Thorpe wrote, "we were able"*: Letter from Philip Thorpe to Reginald Pound, March 11, 1963, 8, Letters to Reginald Pound.

165 *"[t]hese wizards of surgery"*: "Faces Rebuilt. New Hospital to Transform Ugliness into Good Looks. Shattered Men Remade," *Daily Sketch*, July 1917. Found in the Queen's Hospital, Sidcup, Kent: Newspaper Cuttings, London Metropolitan Archive, H02/QM/Y/01/005, page 14.

165 *"Will you withhold your subscription"*: "Worst Loss of All."

165 *"grievously disfigured . . . hero"*: The Editor, "The Queen's Hospital, Frognal, Sidcup. New Jaws and Noses for Wounded Men," *Kent Messenger*, August 1917. Found in the Queen's Hospital, Sidcup, Kent: Newspaper Cuttings, London Metropolitan Archive, H02/QM/Y/01/005, page 16.

165 *"No plastic unit is good"*: Gillies and Millard, *Principles and Art of Plastic Surgery*, 31.

166 *"By merely picking up a phone"*: Ibid.

166 *"the Somme multiplied and intensified"*: Frederick W. Noyes, *Stretcher-Bearers . . . at the Double!* (Toronto: Hunter-Rose, 1937), 177.

166 *"[t]he whole earth is ploughed"*: Quoted in van Bergen, *Before My Helpless Sight*, 90.

167 *"the groans and wails of wounded men"*: Edwin Campion Vaughan, *Some Desperate Glory: The World War I Diary of a British Officer, 1917* (Barnsley, UK: Pen & Sword Military, 2010), 228.

167 *"rats were getting out"*: Quoted in Tim Lynch, *They Did Not Grow Old: Teenage Conscripts on the Western Front, 1912* (Stroud, UK: Spellmount, 2013), 212.

167 *Because of his experiences*: Claire Chatterton and Marilyn McInnes, "'Rekindling the Desire to Live.' Nursing Men Following Facial Injury and Surgery During the First World War," *Bulletin of the UK Association for the History of Nursing* (2016): 57.

168 *"Often have I picked up"*: Private J. McCauley, Imperial War Museums Documents 97/10/1. Originally quoted in Lynch, *They Did Not Grow Old*, 212.

168 *"I shuddered as my hands"*: Ibid.

168 *The men gathered up*: Chatterton and McInnes, "'Rekindling the Desire to Live,'" 58.

169 *He would become Gillies's personal chauffeur*: Bamji, *Faces from the Front*, 192.

169 *When Gillies forgot to renew*: Letter from Allen Daley to Reginald Pound, February 3, 1963, 2, Letters to Reginald Pound.

170 *It was love at first sight*: Beldam's wife would remain by his side until he died from cancer in 1978—over sixty years after he was given a prognosis of six months. Chatterton and McInnes, "'Rekindling the Desire to Live,'" 58.

170 *"I would draw Major Gillies' attention"*: J. L. Aymard, "The Tubed Pedicle in Plastic Surgery," *The Lancet* (July 31, 1920): 270.

170 *This was a rhinoplasty case*: J. L. Aymard, "Nasal Reconstruction. With a Note on Nature's Plastic Surgery," *The Lancet* (December 15, 1917): 888–92.

170 *"I do not intend to enter"*: Aymard, "The Tubed Pedicle in Plastic Surgery," 270.

170 *"operating books, surgical records"*: H. D. Gillies, "The Tubed Pedicle in Plastic Surgery," *The Lancet* (August 7, 1920): 320.

171 *"I must admit that it was"*: Gillies and Millard, *Principles and Art of Plastic Surgery*, 44.

172 *In this sense, the tubed pedicle*: Klaas W. Marck, Roman Palyvoda, Andrew Bamji, and Jan J. van Wingerden, "The Tubed Pedicle Flap Centennial: Its Concept, Origin, Rise and Fall," *European Journal of Plastic Surgery* (February 2017): 473–78.

172 *"It is horrible to feel"*: Letter from Harold Gillies to Sir Squire Sprigge (May 3, 1935). Quoted in Pound, *Gillies*, 109.

172 *In fact, it had long been rumored*: Pound, *Gillies*, 109. See also Meikle, *Reconstructing Faces*, 83.

172 *"I am very sorry"*: Letter from Harold Gillies to J. L. Aymard (January 21, 1939). Quoted in Pound, *Gillies*, 127.

173 *"I, and I think rightly"*: Ibid.

10. PERCY

175 *Private Percy Clare of the 7th Battalion*: Private Papers of P. Clare, vol. 3.

175 *In his hand was a small Bible*: Letter from Percy Clare to his mother (n.d.). Private Papers of P. Clare, Letters to His Mother.

175 *"[O]ur journey was most perilous"*: Private Papers of P. Clare, vol. 3.

176 *"I remember them climbing"*: Ibid.

176 *"took every precaution to prevent"*: Ibid.

176 *Only later would he learn*: Ibid.

176 *"The driver rushed us off"*: Ibid.

177 *"white-gowned surgeons stand so thick"*: La Motte, "The Backwash of War," 32.

177 *"Yes, Sir: through and through"*: Private Papers of P. Clare, vol. 3.

177 *"I certainly couldn't have told them"*: Ibid.

177 *Only later would Clare learn*: Ibid.

177 *"They merely read the label"*: Ibid.

177 *Despite the severity of Clare's injuries*: Malcolm Vivian Hay, *Wounded and a Prisoner of War by an Exchanged Officer* (New York: George H. Doran, 1917), 229–30. Originally found in Bamji, *Faces from the Front*, 38–39.

178 *"What the knowledge that I"*: Private Papers of P. Clare, vol. 3.

178 *"My bristly beard was about"*: Ibid.

178 *A few days later*: John H. Plumridge, *Hospital Ships and Ambulance Trains* (London: Seeley, Service, 1975), 37–39.

179 *Technically, hospital ships were protected*: Ibid., 42–43.

179 *"All the deck machinery fell"*: Quoted in Elizabeth Gleick and Anthee Carassava, "Deep Secrets," *Time International* (South Pacific Edition) 43 (October 26, 1998): 72.

180 *Percy Clare, at least, made it safely*: Private Papers of P. Clare, vol. 3.

180 *"I saw tears run down the sister's cheek"*: Ibid.

181 *"Outside a square brick building"*: P. Gibbs, *Now It Can Be Told* (Garden City, NY: Garden City, 1920), 179–80.

181 *The label pinned to Clare's uniform*: Private Papers of P. Clare, vol. 3.

181 *"The Hospital at Frensham"*: Ibid.

181 *"vinegar faced old 'cat'"*: Ibid.

181 *"[T]he men had no entertainment"*: Ibid.

182 *"There was no provision"*: Ibid.

182 *"The MO [Medical Officer] wanted me"*: Ibid.

182 *"He told me boastingly"*: Ibid.

183 *"Had I not been wounded"*: Ibid.

183 *"Imagine my feelings at the thought"*: Ibid.

183 *By the time he reached the platform*: Ibid.

184 *Between 1915 and 1919*: "'Soldiers' and Sailors' Free Buffet' at Victoria Station," Imperial War Museums, accessed November 16, 2020, https://www.iwm.org.uk/collections/item/object/30019570.

184 *"What kind friends the Tommy ever found"*: Private Papers of P. Clare, vol. 3.

184 *The Queen's Hospital was brimming*: "A Christmas Wonder Tale. How They Spend Yule in the Military Hospitals," *Pall Mall Gazette*, December

24, 1917. Found in the Queen's Hospital, Sidcup, Kent: Newspaper Cuttings, London Metropolitan Archive, H02/QM/Y/01/005, page 26.

184 *Men came round the bed*": Private Papers of P. Clare, vol. 3.

184 *The dark-green enameled beds*: "The Queen's Hospital, Sidcup, The Treatment of Facial and Jaw Injuries," *Nursing Mirror and Midwives' Journal* (August 4, 1917): 309. Found in the Queen's Hospital, Sidcup, Kent: Newspaper Cuttings, London Metropolitan Archive, H02/QM/Y/01/005, page 20. See also Private Papers of P. Clare, vol. 3.

184 *At the foot of each cot*: Private Papers of P. Clare, vol. 3.

184 *A nurse led him to his bed*: Ibid.

185 *"entered into their work"*: "My Personal Experiences of the Great War," 6 Mss Essays by Patients with Facial Injuries in Sidcup Hospital, 1922. LIDDLE/WW1/GA/WOU/34, Essay 4. Liddle Collection, Brotherton Library Special Collections, University of Leeds.

185 *"there could not have been"*: Quoted in Bamji, *Faces from the Front*, 149.

185 *"I lay staring at the shaded light"*: Private Papers of P. Clare, vol. 3.

185 *"They are a cheerful crowd"*: Letter from Percy Clare to his mother (January 8, 1918). Private Papers of P. Clare, Letters to His Mother.

185 *His favorite was a young man*: Ibid.

186 *"By removing them from the atmosphere"*: "The Queen's Hospital, Sidcup. The Treatment of Facial and Jaw Injuries."

186 *Besides leisure activities*: Bamji, *Faces from the Front*, 154.

186 *"extensive gardens where the patients"*: "New Military Queen's Hospital at Frognal, Sidcup, Kent," *The Citizen*, August 4, 1917, 4.

186 *The last was especially useful*: "The Queen's Hospital, Sidcup. The Treatment of Facial and Jaw Injuries."

186 *One of the more well-attended courses*: "Soldier Craftsmen. Display of Work by Hospital Patients," *The Times*, December 9, 1919; "Queen Mary and the Elephant," *Pall Mall Gazette,* December 9, 1919. Found in the Queen's Hospital, Sidcup, Kent: Newspaper Cuttings, London Metropolitan Archive, H02/QM/Y/01/005, page 59.

187 *There was even a hospital barber*: Bamji, *Faces from the Front*, 156.

187 *"Sidcup was indeed a paradise"*: Letter from Percy Clare to his mother (n.d.). Private Papers of P. Clare, Letters to His Mother.

187 *"Shall I whisper a secret to you?"*: Letter from Percy Clare to his mother (n.d.), ibid.

188 *"He can't do anything until"*: Letter from Percy Clare to his mother (December 13, 1917), ibid.

188 *"The boys call going under operation"*: Letter from Percy Clare to his mother (n.d.), ibid.

188 *"[T]here is much laughter"*: Letter from Percy Clare to his mother (January 8, 1918), ibid.

188 *"I had to keep swabbing blood"*: Ibid.

189 *"[We] embraced each other"*: Private Papers of P. Clare, vol. 3.

189 *The snow crunched under the soldiers' boots*: Ibid.

190 *"was shot through the knee-joint"*: Gillies, *Plastic Surgery of the Face*, 40–41. Also see Pound, *Gillies*, 53.

190 *The order to return to duty*: Private Papers of P. Clare, vol. 3.

11. HEROIC FAILURES

191 *The convalescing officers watched*: As retold in Captain J. K. Wilson's account of his service on the Western Front, including the battle of Cambrai in 1917 and his experience at Sidcup (written c. 1970), Private Papers of Captain J. K. Wilson, Documents, 12007, 8. Documents and Sound Archives of the Imperial War Museums. Wilson calls it "the common room," but I believe he's referring to the "Convalescent Officer's Sitting Room," which was situated in the mansion.

192 *Late in the war, Captain "Freddie" West*: "RAF Honours First WW1 Pilot to Win the Victoria Cross," accessed September 29, 2020, https://www.raf.mod.uk/news/articles/raf-honours-first-ww1-pilot-to-win-the-victoria-cross/.

193 *He was first sent to a military hospital*: According to Gillies's case 388: "In addition to the left eye being burned and to all the other destruction in evidence, the right eye was practically blind, as a result of staphyloma of the cornea." Gillies, *Plastic Surgery of the Face*, 364.

193 *"His face is burned beyond recognition"*: Letter from Agnes Keyser to Sir Reginald Wilson, February 1917. Quoted in Bamji, *Faces from the Front*, 24.

193 *"it had to be decided"*: Gillies, *Plastic Surgery of the Face*, 364.

194 *"It has been a good puff"*: Millard, "Gillies Memorial Lecture," 76.

194 *Before every major operation*: Pound said that Gillies did this before every major operation. Pound, *Gillies*, 51. For more on the organization of the hospital, see "The Queen's Hospital, Frognal, Sidcup," *The Lancet* (November 3, 1917): 687–89.

194 *"Can our general surgeons"*: Gillies and Millard, *Principles and Art of Plastic Surgery*, 46.

194 *"It is impossible at times"*: Ibid., 50.

195 *Gillies later described the process*: Gillies, *Plastic Surgery of the Face*, 364.

195 *"the patient was considerably collapsed"*: Ibid.

195 *"Both the chest area"*: Ibid.

195 *"to obtain a perfect result"*: Ibid.

195 *"One could have wished"*: Ibid.

196 *"In the trenches at night"*: "Voices of the First World War: The German Spring Offensive," Imperial War Museums podcast, accessed July 5,

2021, https://www.iwm.org.uk/history/voices-of-the-first-world-war
-the-german-spring-offensive.

196 *"Never let routine methods"*: Gillies and Millard, *Principles and Art of Plastic Surgery*, 53.

196 *Henry Tonks loomed*: Lindsay, "Five Men," 63.

196 *The Australian artist*: "Daryl Lindsay: Late in Life an Old Dream Is Coming True," *The Age*, August 4, 1962, 18.

197 *Lindsay had been working*: Lindsay, "Five Men," 62.

197 *"I can see him now"*: Ibid.

197 *Lindsay introduced himself to the surgeon*: Ibid.

197 *Afterward, Lindsay met with Newland*: Ibid.

198 *"Trying to draw"*: Ibid., 63.

198 *"I'm glad you said 'trying'"*: Ibid.

198 *And so it transpired*: Ibid.

198 *"Not a small feature"*: Gillies, *Plastic Surgery of the Face*, x–xi.

199 *There were also photographers*: Bamji, *Faces from the Front*, 123.

199 *The earliest medical photograph*: Emily Milam, "A Brief History of Early Medical Photography," *Clinical Correlations*, September 30, 2016, accessed December 22, 2020, https://www.clinicalcorrelations.org/2016/09/30/a-brief-history-of-early-medical-photography/.

200 *Despite the presence of photographers*: Pound, *Gillies*, 159.

200 *"[H]e was as full of ideas"*: Quoted ibid.

200 *"Oh God, this is the end"*: "Voices of the First World War: The German Spring Offensive."

201 *"Private Bell is a very fine chap"*: Quoted in Gillies and Millard, *Principles and Art of Plastic Surgery*, 15.

201 *But this particular letter worried Gillies*: Pound, *Gillies*, 50. Pound says that Gillies was "sweating with fear" over what to do about Bell's case. See also Gilles and Millard, *Principles and Art of Plastic Surgery*, 15.

201 *After Gillies returned to Britain*: Joseph Harbison, "The 13th Stationary/83rd (Dublin) General Hospital, Boulogne, 1914–1919," *Journal of the Royal College of Physicians of Edinburgh* 45 (2015): 229–35.

201 *During the final years of the war*: McAuley, "Charles Valadier: A Forgotten Pioneer," 785.

201 *Before Thorpe was shipped off*: Meikle, *Reconstructing Faces*, 49.

202 *"[H]e offered to do the job himself"*: Letter from Philip Thorpe to Reginald Pound, March 11, 1963, 5–6, Letters to Reginald Pound.

202 *"His little mouth opened vertically"*: Gillies and Millard, *Principles and Art of Plastic Surgery*, 15.

202 *"Obviously the patient could not remain"*: Ibid.

203 *A nurse presented him with a linen bag*: Pound, *Gillies*, 54.

203 *"horrid blue colour"*: Gillies, *Plastic Surgery of the Face*, 87.

204 *"The first step toward filling"*: Gillies and Millard, *Principles and Art of Plastic Surgery*, 16.

204 *"I was more than thankful"*: Gillies, *Plastic Surgery of the Face*, 87.

204 *"A good style will get you through"*: Quoted in Bamji, *Faces from the Front*, 134.

12. AGAINST ALL ODDS

205 *Tapers of sunlight*: Pound, *Gillies*, 56.

205 *While the challenges facing Gillies*: J. M. McDonald, "Anaesthesia on the Western Front—Perspectives a Century Later," *Anaesthesia and Intensive Care* 44 Suppl. (2016): 16.

206 *Over the course of the war*: W. G. MacPherson, *Medical Services General History*, vol. 1 (London: HMSO, 1921), 180. See also N. H. Metcalfe, "The Effect of the First World War (1914–1918) on the Development of British Anaesthesia," *European Journal of Anaesthesiology* 24, no. 8 (2007): 649–57.

206 *"I spent most of my time"*: Quoted in McDonald, "Anaesthesia on the Western Front," 18.

207 *"the surgeon must perforce trespass"*: Gillies, *Plastic Surgery of the Face*, 23.

207 *"[W]hen a boy was notified of an operation"*: Gillies and Millard, *Principles and Art of Plastic Surgery*, 57.

207 *"the sight of a man in a white coat"*: Pound, *Gillies*, 32.

207 *"Positive pressure was necessary"*: Gillies and Millard, *Principles and Art of Plastic Surgery*, 60.

208 *In 1919, Magill and his team*: Peter Bodley, "Development of Anaesthesia for Plastic Surgery," *Journal of the Royal Society of Medicine* 71 (November 1978): 842.

208 *Just as Gillies promoted the cause*: For more on Magill and his contributions to anesthesia, see Bamji, *Faces from the Front*, 109–10.

208 *Private Stanley Girling, who sustained*: "Residents Who Served, Girling, Stanley (Gunner)," accessed February 11, 2021, https://www.saanich.ca/EN/main/parks-recreation-culture/archives/saanich-remembers-wwi/residents-who-served-a-l.html.

209 *In the early twentieth century*: Carl Zimmer, "Why Do We Have Blood Types?," *BBC Future*, July 15, 2014, accessed February 17, 2020, https://www.bbc.com/future/article/20140715-why-do-we-have-blood-types.

209 *So, he collected blood*: Ibid.

210 *In March 1914—just four months*: Luis Agote is sometimes credited as performing the first transfusion of citrated blood. However, Agote performed his transfusion on November 9, 1914, nearly eight months after Hustin performed his own citrated transfusion.

210 *"This great stride forward"*: Geoffrey Keynes, *Blood Transfusions* (Oxford: Oxford Medical Publications, 1922), 17.

210 *The first soldier to receive a transfusion*: F. Boulton and D. J. Roberts, "Blood Transfusion at the Time of the First World War—Practice and Promise at the Birth of Transfusion Medicine," *Transfusion Medicine* 24 (2014): 329. See also Rose George, *Nine Pints: A Journey Through the Mysterious, Miraculous World of Blood* (London: Portobello Books, 2018), 75–77.

211 *In 1916, a surgeon named Andrew Fullerton*: A. Fullerton, G. Dreyer, and H. C. Bazett, "Observations on Direct Transfusion of Blood, with a Description of a Simple Method," *The Lancet* (May 12, 1917): 715–19. Despite the high failure rate, Fullerton felt his results were good, since all the cases were desperate. Saving two lives was better than saving none, even if it meant that fifteen others still died.

211 *The truth was that the existing techniques*: S. L. Wain, "The Controversy of Unmodified Versus Citrated Blood Transfusion in the Early 20th Century," *Historical Review* 24, no. 5 (1984): 405.

211 *"ought to be used much more widely"*: Fullerton, Dreyer, and Bazett, "Observations on Direct Transfusion," 715.

211 *It wasn't until 1917*: J. R. Hess and P. J. Schmidt, "The First Blood Banker: Oswald Hope Robertson," *Transfusion* 40, no. 1 (2000): 110–13.

212 *"Being pitched into a hospital's service"*: Letters to Peyton Rous, O. H. Robertson's Papers, American Philosophical Society, Philadelphia, PA, dated June 27, 1917, quoted in William C. Hanigan and Stuart C. King, "Cold Blood and Clinical Research During World War I," *Military Medicine* 161, no. 7 (1996): 394.

212 *"difficulty of procuring sufficient blood"*: Oswald H. Robertson, "Transfusion with Preserved Red Blood Cells," *British Medical Journal* 1 (1918): 691–95.

212 *"blood on tap"*: Letters to Peyton Rous, O. H. Robertson's Papers, American Philosophical Society, Philadelphia, PA, dated June 27, 1917, quoted in Hanigan and King, "Cold Blood and Clinical Research During World War I," 394.

212 *"[it] was then that I realized"*: Letters to Peyton Rous, O. H. Robertson's Papers, American Philosophical Society, Philadelphia, PA, dated December 29, 1917, quoted in Hanigan and King, "Cold Blood and Clinical Research During World War I," 395.

213 *He was able to convince a colonel*: Hanigan and King, "Cold Blood and Clinical Research During World War I," 395.

213 *One resuscitation team was led*: Boulton and Roberts, "Blood Transfusion at the Time of the First World War," 31.

213 *"To the serious scientist"*: Quoted in Hanigan and King, "Cold Blood and Clinical Research During World War I," 397–98.

214 *Newspapers declared that the young man*: "Brother Who Gave His Life," *Sunday Pictorial*, October 20, 1918, 1.

214 *By July 1918, Henry Tonks*: For more, see Julian Freeman, "Professor Tonks: War Artist," *Burlington Magazine* 127, no. 986 (May 1985): 284–93.

215 *"a peculiarly vicious gun"*: Tonks to Yockney, July 25, 1918; Imperial War Museums, Tonks correspondence file, quoted ibid., 289.

216 *"I have seen enough"*: Tonks to Yockney, September 14, 1918, Imperial War Museums, Tonks correspondence file, quoted ibid., 290.

216 *From then on, it became widely known*: Laura Spinney, *Pale Rider: The Spanish Flu of 1918 and How It Changed the World* (New York: PublicAffairs, 2017), 151–63.

217 *Unsurprisingly, epidemics have also traveled*: For a lengthy discussion of this subject, see Mark Osborne Humphries, "Paths of Infection: The First World War and the Origins of the 1918 Influenza Pandemic," *War in History* 21, no. 1 (January 2014): 55–56.

218 *"trail of infected armies"*: Colonel Guy Carleton Jones, "The Importance of the Balkan Wars to the Medical Profession of Canada," *Canadian Medical Association Journal* 4, no. 9 (1914): 801–802.

218 *"[o]ne of the most striking of the complications"*: George R. Callender and James F. Coupal, *The Medical Department of the United States Army in the World War: Pathology of the Acute Respiratory Diseases, and of Gas Gangrene Following War Wounds*, Vol. 12 (Washington, DC: U.S. Government Printing Office, 1929), 57.

218 *It was said that a person*: This story may be apocryphal.

218 *"It is only a matter of a few hours"*: N. R. Grist, "Pandemic Influenza 1918," *British Medical Journal* (December 2, 1979): 1632–33.

219 *Nurse Buckler was among them*: Bexley Borough WW1 Roll of Honour, accessed July 9, 2021, https://www.bexley.gov.uk/discover-bexley/archives-and-local-history/local-history-resources/bexley-remembers-first-world-war.

13. ALL THAT GLITTERS

221 *Daryl Lindsay was making his way up a ramp*: Lindsay, "Five Men," 62.

221 *"surging mass" of people who "wandered aimlessly"*: *Daily Mirror*, November 14, 1918, 2.

222 *Exhumations continued with regularity*: van Bergen, *Before My Helpless Sight*, 493.

222 *"[A]s we spoke," Lindsay observed*: Lindsay, "Five Men," 62.

222 *Later, when reflecting on the moment*: Gillies and Millard, *Principles and Art of Plastic Surgery*, 43.

223 *The days may have been growing colder*: As told in Reginald Pound, *Gillies*, 55.

224 *Five years earlier, Jugon*: Shane A. Emplaincourt, "La Chambre des Officiers and Recapturing the Evanescent Memory of the Great War's Gravely Disfigured," *War, Literature & the Arts* 30 (2018): 18.

225 *"[I]t was amusing to notice"*: The Times, June 30, 1919, 13.

225 *"Nearly everyone seemed to have a camera"*: *Lancashire Daily Post*, June 30, 1919, 2.

225 *The German representatives*: Ibid.

226 *"ceremony was curiously unimpressive"*: Ibid.

226 *Far from the grandeur of Versailles*: Millard, "Gillies Memorial Lecture," 76.

EPILOGUE: CUTTING A PATH

229 *In the spring of 1920*: Bamji, *Faces from the Front*, 161.

230 *In 1925, the eight remaining facial patients*: Ibid., 159.

230 *"With tears in his eyes"*: Pound, *Gillies*, 68.

231 *Valadier was one of only two dentists*: Cruse, "Auguste Charles Valadier: A Pioneer in Maxillofacial Surgery," 337–38.

231 *The disfigured face remained*: For a thoughtful critique, see Biernoff, "The Rhetoric of Disfigurement," 666–85.

231 *"[i]t horrified Tonks to find"*: Hone, *The Life of Tonks*, 175.

232 *"I have loved my students"*: Quoted ibid., 224–25.

232 *"[M]y painting is more than my amusement"*: Quoted ibid., 230.

232 *"it was not like Sidcup"*: Private Papers of P. Clare, vol. 3.

232 *"I handed in my khaki suit"*: Ibid.

233 *"[T]o my amazement, such monetary"*: Quoted in Layton, *Sir William Arbuthnot Lane*, 110.

233 *"You won't be writing"*: Mick Gillies, *Mayfly on the Stream of Time*, 2.

233 *"Dear Facemaker"*: Quoted in Pound, *Gillies*, 93.

233 *"a personal honour but as one shared"*: Quoted ibid., 97.

234 *Shortly after the public announcement*: Ibid., 93–94.

234 *"To venture into this rather new field"*: Gillies and Millard, *Principles and Art of Plastic Surgery*, 391.

234 *"It meant reassociation"*: Ibid.

235 *"Don't be a fool"*: Quoted in Pound, *Gillies*, 64.

235 *"the time may yet hardly"*: "Plastic Surgery of the Face," *The Lancet* (July 24, 1920): 194.

235 *He was determined to prove*: Gillies and Millard, *Principles and Art of Plastic Surgery*, 391.

235 *"Name plate up. Secretary installed"*: Quoted in Pound, *Gillies*, 66.

235 *"Reconstructive surgery is an attempt"*: Gillies and Millard, *Principles and Art of Plastic Surgery*, 395.

235 *The work of Jacques Joseph*: Samuel M. Lam, "John Orlando Roe: Father of Aesthetic Rhinoplasty," *Archives of Facial Plastic Surgery* 4 (April–June 2002): 122–23. See also Elizabeth Haiken, "The Making of the Modern Face: Cosmetic Surgery," *Social Research* 67, no. 1 (2000): 81–97; and Michelle Smith, "The Ugly History of Cosmetic Surgery," *The Independent*, June 10, 2016.

236 *Chief among them was Charles Conrad Miller*: John B. Mulliken, "Biographical Sketch of Charles Conrad Miller, 'Featural Surgeon,'" *Plastic and Reconstructive Surgery* 59 (February 1977): 175–84.

236 *"upon honest effort in the uplifting"*: "Publications. Cosmetic Surgery. *The Correction of Featural Imperfections* by Charles C. Miller," *California State Journal of Medicine* 6, no. 7 (July 1908): 244–45.

236 *"It is easier to reduce"*: Gillies and Millard, *Principles and Art of Plastic Surgery*, 395.

236 *"Hugo was for the Grecian type"*: Ibid., 427.

237 *"These things take time"*: Quoted in Pound, *Gillies*, 58.

237 *"I have since been informed"*: Gillies and Millard, *Principles and Art of Plastic Surgery*, 391.

237 *Gillies was always able to find the humor*: As retold in Captain J. K. Wilson's account of his service on the Western Front, including the Battle of Cambrai in 1917 and his experience at Sidcup (written c. 1970). Private Papers of Captain J. K. Wilson. Documents. 12007, 81–82. Documents and Sound Archives of the Imperial War Museums.

238 *"I do plead guilty to casting"*: Gillies and Millard, *Principles and Art of Plastic Surgery*, 428.

238 *"Sir Harold's face-lifting operations"*: Letter from Frances Steggall to Reginald Pound. Letters to Reginald Pound.

238 *"As there seemed to be some time"*: Quoted in Pound, *Gillies*, 86.

239 *"Without looking into his mouth"*: Quoted ibid.

239 *"would give a young assistant"*: Ibid., 130.

239 *"Talk it over with my secretary"*: Gillies and Millard, *Principles and Art of Plastic Surgery*, 427.

239 *"woman who has children"*: Ibid.

239 *"Every effort, however, should be made"*: Ibid.

239 *"I am afraid your ideas"*: Quoted in Pound, *Gillies*, 129.

240 *"[I]f you are not going to be vain"*: Gillies and Millard, *Principles and Art of Plastic Surgery*, 425.

240 *"Often while lifting a face"*: Ibid., 395.

240 *"little extra happiness [it brings]"*: Ibid.

NOTES

240 *While many clients looked*: Virat Markandeya, "When Deadly X-Rays Were Used for Hair Removal," *Ozy*, November 26, 2019, accessed October 20, 2020, https://www.ozy.com/true-and-stories/when-hair-removal -was-a-public-health-crisis/220770/. See also Rebecca Herzig, *Plucked: A History of Hair Removal* (New York: New York University Press, 2015).

240 *"While I'm certain that I removed"*: Quoted in Pound, *Gillies*, 128.

241 *"most of the flesh"*: Letter from Mrs. V.F.E. Gerrard to Reginald Pound. Letters to Reginald Pound.

241 *"Burns received in war"*: Gillies and Millard, *Principles and Art of Plastic Surgery*, 445.

241 *"What a harrowing decision"*: Ibid.

242 *"But where to begin"*: Ibid.

242 *"to break the monotony"*: Ibid., 446.

242 *"In all tube pedicle mouths"*: Ibid.

242 *Before she left the hospital*: Pound, *Gillies*, 82.

242 *"She was never again"*: Letter from Mrs. V.F.E. Gerrard to Reginald Pound. Letters to Reginald Pound.

242 *"Throughout her entire reconstruction"*: Gillies and Millard, *Principles and Art of Plastic Surgery*, 446.

243 *"Poor Mrs. Brown"*: Quoted in Pound, *Gillies*, 82.

243 *"All that can be seen now"*: Letter from Mrs. V.F.E. Gerrard to Reginald Pound. Letters to Reginald Pound.

243 *"[C]onsidering that Sir Harold"*: Ibid.

243 *"If you were starting this case"*: Pound, *Gillies*, 164.

243 *"The disappointment when Morestin"*: Gillies and Millard, *Principles and Art of Plastic Surgery*, 392.

244 *"young eager minds"*: Ibid.

244 *Peering out from among the rubble*: Reports indicate that Tonks's portraits were in the building at the time of the bombing, though it is possible that they had been removed beforehand. A.J.E. Cave, "Museum," *Royal College of Surgeons of England. Scientific Report* (1940–1941): 4, 10.

245 *Gillies's talents and hard-won skills*: For more on Archibald McIndoe, see Emily Mayhew, *The Reconstruction of Warriors: Archibald McIndoe, the Royal Air Force and the Guinea Pig Club* (London: Greenhill Books, 2004).

245 *"[Y]ou know it made my blood boil"*: Letter from Horace Sewell to Reginald Pound. Letters to Reginald Pound.

245 *"I had taken her seriously"*: Michael Dillon and Lobzang Jivaka, *Out of the Ordinary: A Life of Gender and Spiritual Transitions*, eds. Jacob Lau and Cameron Partridge (New York: Fordham University Press, 2017), 89–90. See also Brandy Schillace, "The Surprisingly Old Science of Living as Transgender," *Scientific American* (March 18, 2020).

245 *But Dillon wished to complete his transition*: Gillies and Millard, *Principles and Art of Plastic Surgery*, 379; Dillon and Jivaka, *Out of the Ordinary*, 8.

246 *"The world began to seem"*: Dillon and Jivaka, *Out of the Ordinary*, 102.

246 *Over a span of several years*: Ibid., 104. See also Andrew N. Bamji and Peter J. Taub, "Phalloplasty and the Tube Pedicle: A Chronological Re-evaluation," *European Journal of Plastic Surgery* 43 (2020): 7–12.

246 *By rolling a tube of tissue*: Rajesh Nair, "Sir Harold Gillies: Pioneer of Phalloplasty and the Birth of Uroplastic Surgery," *Journal of Urology* 183 (May 31, 2010): e437.

246 *In 1949, Gillies became*: Karl Baer is sometimes erroneously identified as the first trans man to undergo phalloplasty. Baer was intersex and was born with hypospadias, a comparatively common birth defect resulting in the displacement of the urethra on the penis. As a result, he was misgendered at birth and raised as a girl. As an adult, he visited the Institute for Sexual Research, led by eminent German sexologist Magnus Hirschfeld. After undergoing an examination, Baer was allowed to change his sex legally. It is unclear whether he underwent any surgical procedure, since the case notes have since been destroyed or lost. For more on Baer, see J. Funke, "The Case of Karl M.[artha] Baer: Narrating 'Uncertain' Sex," in *Sex, Gender and Time in Fiction and Culture*, eds. B. Davies and J. Funke (London: Palgrave Macmillan, 2011), 132–53.

246 *"How different was life now"*: Dillon and Jivaka, *Out of the Ordinary*, 109.

246 *"He always seemed glad"*: Ibid., 187.

246 *In 1958, British journalists*: When Dillon was outed as a trans man by British journalists in 1958, Gillies wrote to his former patient, offering his support. "Letters started coming now from my oldest friends, offering their sympathy and saying what they thought of the press. Sir Harold Gillies also wrote, and the Lady Warden from the Mission to Seamen, and, of course, Lobzang Rampa, who himself had been in the papers in the past two months. One and all they wrote encouragement." Dillon and Jivaka, *Out of the Ordinary*, 217.

247 *"one aim had always been"*: Ibid., 102.

247 *"no evil ever befalls"*: Albee, *A Surgeon's Fight to Rebuild Men*, 134.

247 *"one day surgeons will do something"*: Gillies and Millard, *Principles and Art of Plastic Surgery*, 629.

247 *"We have been placing high odds"*: Millard, "Gillies Memorial Lecture," 77.

248 *"never do today what you can put off "*: Ibid., 78.

248 *"proud to have received them"*: Quoted in Pound, *Gillies*, 225.

248 *Some plastic procedures, such as rhinoplasty*: Meikle, *Reconstructing Faces*, 81.

NOTES

248 *"As a result of [the] efforts"*: Neal Owens, "To Sir Harold Gillies," *American Journal of Surgery* 95, no. 2 (February 1958): 167.

249 *The procedure, which is considered*: Fay Bound Alberti and Victoria Hoyle, "Face Transplants: An International History," *Journal of the History of Medicine and Allied Sciences* 76, no. 3 (July 2021): 319–45, accessed July 9, 2021, https://doi.org/10.1093/jhmas/jrab019.

ACKNOWLEDGMENTS

Although I had always been fascinated by Harold Gillies and the First World War, *The Facemaker* was not the book I intended to write after *The Butchering Art*. I worried that it was too big a subject to tackle and that I was not the best-placed person to write this story. I pitched several ideas to my publisher, but as fate would have it, this is the one that stood out to them. I went into this project on a wing and a prayer. Had it not been for some encouraging words from my friend Erik Larson, I may never have had the confidence to move forward with it.

In the end, I'm glad that I did. But I could not have done it without support from a tremendous number of people along the way.

First, I'd like to thank Professor Mark Harrison, Dr. Tim Cook, Dr. Adam Montgomery, Dr. Catherine Kelly, Dr. Paul Schofield, and Bejamin Palmer for providing invaluable feedback on early drafts of *The Facemaker*. Your expert insights coupled with your enthusiasm for this story have made this a better book.

I'd also like to thank the disability activist and author Ariel Henley. As a person living with a facial disfigurement, you offered a unique perspective on *The Facemaker*. Your insights and comments helped me contextualize the experiences of Harold Gillies's patients in ways that would have been impossible without your help, and I am hugely grateful to you for your careful feedback on the manuscript.

ACKNOWLEDGMENTS

When I sat down to write *The Facemaker*, I knew I wanted to drop the reader right into the middle of the action from page one. I never would have been able to do this without the help of Rachel Gray, the great-niece of Percy Clare, who kindly granted me permission to use his diary so that I might piece together the chaotic events that led to his injuries. If Gillies is the backbone of this story, Clare (and the other men who were injured) is its beating heart.

I'd also like to thank Dr. Andrew Bamji, the Gillies Archivist, who discovered thousands of clinical records from World War I while working as Director of Medical Education at Queen Mary's Hospital, Sidcup. Had it not been for his efforts to preserve this material, many of the stories told in this book would have been lost to future generations. His contributions to the subject are a valuable resource.

An author is only as good as her editors, and I was lucky to have several world-class editors working on this book at different stages. First and foremost, I'd like to thank my primary editor at Farrar, Straus and Giroux, Alex Star, whose insights and feedback have made this story richer and more complex. Thank you for keeping me calm during this frenzied process. I'd also like to thank Colin Dickerman for his invaluable guidance at the beginning, when the book was merely an idea. And my UK editor, Laura Stickney, whose comments at the end helped sharpen the narrative.

I also want to thank Devon Mazzone, who manages my foreign rights and never gets irritated by my emails asking when France will buy my book (answer: never). And Ian Van Wye, whose sharp eye helped to streamline the prose and eliminate over a thousand unnecessary words.

I'd like to extend my heartfelt gratitude to my research assistant, Caroline Overy, without whom this book would have taken ten years rather than five. Thank you for helping me navigate the complicated world of copyrights and image clearances and for keeping me organized while I was writing such a complex story. I promise never to take us into the twentieth century again!

Finishing a book during a pandemic is not easy. I am extremely grateful to the many archivists and librarians who facilitated access to their collections despite closures and lockdowns. I am especially grateful to Victoria Rea and the Royal College of Surgeons of England, and to Libby Gavin at BAPRAS. You are the unsung heroes of the research world. The images you have helped me source have added tremendous value to this book.

I'd also like to thank my agent, Robert Guinsler, who came on board this project halfway through it during a tumultuous time in my career. And my manager, Jorge Hinojosa, whose tireless advocacy of my work has bolstered not only my platform, but also my confidence. Thank you for believing in the value of what I do. I will try not to be such a pessimist in the future!

I'm fortunate to have the love and support of countless family members and friends. With that in mind, I'd like to extend my heartfelt thanks to Lucy Coleman Talbot, who has opened my eyes to the world in more ways than one. Thank you for your friendship and for holding me to a higher standard. Not only is this book better for your suggestions, but I am a better person for knowing you.

Thanks to my producing partner and close friend, Lori Korngiebel, who has stuck by me through the highs and lows. And Shelley Estes—my "forever drinking buddy"—who has supported me from the beginning. Thanks also to Monica Walker, who is just as geeky as I am when it comes to medical history.

I'd like to express my gratitude to Kate So, who has been such a supportive friend in the last few years. You ask nothing of anyone, while being incredibly generous to the ones you love. I'm very lucky to have you in my life.

Thanks to Lucy Campbell, my "Snuggly Gorilla," who shows me every day what it means to be a good and loyal friend. And

Dave Brown, who has worked hard, despite my protestations, to ensure that I'm fit enough to endure a book tour without back pain. A very special thanks to my true blue, Estelle Paranque, for all the love, laughter, and support through the years.

I would like to pay tribute to my dear friend Bill MacLehose, who died suddenly in 2020. He believed in me in those early days when I was just starting out as a writer, and I miss him every day.

I'd like to thank my mom and stepdad, Debbie and Greg Klebe, who are always there to pick me up when I am down. And my in-laws, Graham and Sandra Teal, who have cheered me on this past year when I found the isolation from the pandemic almost unbearable. Also, thanks to my brother, Chris Fitzharris, whose sarcasm keeps me grounded (#ForeverBarb). And my dad and stepmom, Mike and Sue Fitzharris, who have traveled the country to support my literary dreams.

My eternal thanks go to my grandmother Dorothy Sissors, who has shown herself to be a bonus in all her grandchildren's lives—especially mine.

And last, but certainly not least, I'd like to thank my husband, Adrian Teal. You are the first person to read what I write and the last person to comment on a manuscript before it goes to the publisher. I value your opinion above all others. I could not do this without you at my side. I love you with all my heart.

INDEX